TELEWORK:
PRESENT SITUATION AND FUTURE DEVELOPMENT OF A NEW FORM OF WORK ORGANIZATION

TELEWORK:
PRESENT SITUATION AND FUTURE DEVELOPMENT OF A NEW FORM OF WORK ORGANIZATION

edited by

Werner B. KORTE
Simon ROBINSON
Wolfgang J. STEINLE

*Empirica GmbH
Bonn, Federal Republic of Germany*

1988

NORTH-HOLLAND
AMSTERDAM · NEW YORK · OXFORD · TOKYO

©ELSEVIER SCIENCE PUBLISHERS B.V., 1988

All rights reserved. No part of this publication may be reproduced, stored in a retrieval system, or transmitted, in any form or by any means, electronic, mechanical, photocopying, recording or otherwise, without the prior permission of the copyright owner.

ISBN: 0 444 70355 1

Published by:
ELSEVIER SCIENCE PUBLISHERS B.V.
P.O. Box 1991
1000 BZ Amsterdam
The Netherlands

Sole distributors for the U.S.A. and Canada:
ELSEVIER SCIENCE PUBLISHING COMPANY, INC.
52 Vanderbilt Avenue
New York, N.Y. 10017
U.S.A.

LIBRARY OF CONGRESS
Library of Congress Cataloging-in-Publication Data

```
Telework : present situation and future development of a new form of
work organization / edited by Werner B. Korte, Simon Robinson,
Wolfgang J. Steinle.
      p.  cm.
   Based on papers from an international conference sponsored by the
Commission of the European Communities and the German Federal
Ministry of Research and Technology held at the
Wissenschaftszentrum, Bonn, Mar. 18-20, 1987.
   Bibliography: p.
   ISBN 0-444-70355-1
   1. Telecommuting--European Economic Community countries-
-Congresses.  2. Telecommuting--Congresses.   I. Korte, Werner B.
II. Robinson, Simon, 1953-    .  III. Steinle, Wolfgang J., 1953-
IV. Commission of the European Communities.  V. Germany (West).
Bundesministerium für Forschung und Technologie.
HD2336.E86T45 1988
331.25--dc19                                               87-30898
                                                               CIP
```

PRINTED IN THE NETHERLANDS

PREFACE AND ACKNOWLEDGEMENTS

The use of electronic media and data processing equipment is altering the structure, routine and contents of office work. Innovations in the information and communication technologies have brought forth a novel form of work organization which has recently received significant interest: telework. The increased implementation of telework would imply extensive structural changes in companies and at the work place. By linking office automation with telecommunications, telework would allow considerable decentralization of office work, with regard to both location and organizational control. Once completed, work could then be electronically transmitted to a given receiving station by means of existing communication networks. However, despite the decentralizing potential of telework, this form of work can, and often does, lead to the centralization of decision structures.

The level of interest in, the acceptance of and satisfaction with telework of both those working with it and their employees vary considerably, as do the efficiency of telework implementations, the qualifications they require, as well as the impact they have on the labour market. This variability is not due merely to organizational or national differences. The numerous predictions concerning the development and diffusion of the various forms of telework are contradictory in their conclusions. Some see in it the future of work organization and predict its sweeping advancement, whereas others are sceptical about its long-term prospects and feel that due to the considerable organizational effort involved, telework applications will remain limited in scope.

The full range of issues involved in telework was the subject of an international conference on "Telework - Present Situation and Future Development of a New Form of Work Organization" initiated and organized by empirica, a German based research and consultancy company. The conference, sponsored by the Commission of the European Communities and the German Federal Ministry of Research and Technology, was held at the Wissenschaftszentrum, Bonn, March 18 to 20, 1987. The aims of the conference were as follows:

o Overview of the present situation of telework (decentralized information and communication technology based office work) in the EEC and the U.S.A.;

o Discussion of technical, economical, organizational, legal, social and psychological aspects of the various forms of telework;

o Discussion of the likely applicational areas and different organizational forms of telework;

o Identification, analysis and discussion of strategies and instruments for the stimulation of desirable forms of telework.

It brought together the world's experts responsible for the groundbreaking international research and practical experiences on the subject. This volume, which is based on papers from the international conference, presents for the first time in one source current international knowledge on the issue of telework.

We wish to thank the members of the programme committee for their work in assisting in developing and shaping the conference, the chairmen for their work in chairing the sessions, and the speakers for their substantial contributions. Special thanks are due to staff members of empirica, namely Maria Grünhage, Christine Künkel and Horst W. Scharbert for their assistance in organizing the conference and editing the transcript of the conference.

We also wish to extend thanks to our sponsors, who made the conference possible: the Commission of the European Communities and the German Federal Ministry of Research and Technology, and last but not least the participants from so many different countries.

Werner B. Korte   Simon Robinson   Dr. Wolfgang J. Steinle

PROGRAMME COMMITTEE

John Alic
(Office of Technology Assessment, Washington)

Prof. Dr. Niels Bjørn-Andersen
(Copenhagen School of Economics and Business Administration)

Prof. Dr. Hans-Jörg Bullinger
(Fraunhofer-Institut für Arbeitswirtschaft und Organisation IAO, Stuttgart)

Ursula Huws (empirica, London)

Trevor Jones
(Commission of the European Communities, DG XIII, Luxembourg)

Eberhard Köhler
(European Foundation for the Improvement of Living and Working Conditions, Dublin)

Dr. Hans-Edgar Martin
(SIEMENS AG, Unternehmensbereich Kommunikation und Datentechnik, Munich)

Dr. Riccardo Petrella
(Commission of the European Communities, FAST, Brussels)

Jan Roukens
(Commission of the European Communities, ESPRIT, Brussels)

Prof. Dr. Claudio Roveda
(Centro lombardo per lo sviluppo tecnologico e produttivo delle piccole e medie Imprese, CESTEC, Milan)

Gitte Vedel
(Copenhagen School of Economic and Social Research, Institute for Industrial Research and Social Development)

Coordination:

Werner B. Korte
Simon Robinson
Dr. Wolfgang J. Steinle

CONTENTS

Preface and Acknowledgements ... v

Programme Committee ... vii

Introduction ... 1

Welcoming Addresses
Jan Roukens
Commission of the European Communities, Brussels, Belgium ... 3
Bernd Kramer
Federal Ministry of Research and Technology, Bonn, Fed. Rep. Germany ... 5

Telework: Opening Remarks on an Open Debate
Wolfgang J. Steinle
Empirica GmbH, Bonn, Fed. Rep. Germany ... 7

1. PRACTICAL EXPERIENCE
Chairman: Jan Roukens
Commission of the European Communities, Brussels, Belgium

Telework in the UK
"Steve" Shirley
F International Group Plc., Chesham, United Kingdom ... 23

Towards New Patterns of Work
Phil E. Judkins
Rank Xerox Ltd., London, United Kingdom ... 33

The Organizational Development of Teleprogramming
Wolfgang Heilmann
INTEGRATA GmbH, Tübingen, Fed. Rep. Germany ... 39

2. ECONOMIC AND SOCIAL FACTORS
Chairman: Amin Rajan
Institute of Manpower Studies, University of Sussex, United Kingdom

Remote Possibilities: Some Difficulties in the Analysis and
    Quantification of Telework in the UK
Ursula Huws
Empirica UK, London, United Kingdom ... 61

Organizational Barriers to Telework
Margrethe H. Olson
*Graduate School of Business Administration, New York University,
    United States* ... 77

Social Aspects of Telework: Facts, Hopes, Fears, Ideas
Marilyn Mehlmann
Delfi Consult AB, Stockholm, Sweden ... 101

3. ORGANIZATIONAL AND TECHNICAL ASPECTS
Chairman: Herbert Kubicek
University of Trier, Fed. Rep. Germany

The Dilemma of Telework: Technology vs. Tradition
Gil E. Gordon
Gil Gordon Associates, Monmouth Junction, United States — 113

Decentralization Via Teletex: Organizational and Technical Impact Experiences of the Research Project "Creation of Decentralized Work Places through Teletex"
Barbara Klein and Hans-Peter Fröschle
Fraunhofer Institut für Arbeitswirtschaft und Organisation, Stuttgart, Fed. Rep. Germany — 137

4. FUTURE PERSPECTIVES AND STRATEGIES
Chairman: Hans-Jörg Bullinger
Fraunhofer, IAO, Stuttgart, Fed. Rep. Gremany

Telework - Potential, Inception, Operation and Likely Future Situation
Werner B. Korte
Empirica GmbH, Bonn, Fed. Rep. Germany — 159

(Tele-) Homework in the Federal Republic of Germany: Historical Background and Future Perspectives from a Worker's Perspective
Herbert Kubicek and Ulrich Fischer
University of Trier, Fed. Rep. Germany — 177

Autonomy, Telework and Emerging Cultural Values
Gerard Blanc
Association Internationale Futuribles, Paris, France — 189

Trends of Decentralization of White Collar Activitites by Means of Information and Communication Technologies
Helmut Drüke, Günter Feuerstein, and Rolf Kreibich
Institut für Zukunftsstudien und Technologiebewertung, Berlin, Fed. Rep. Gremany — 201

5. TELEWORK FROM A EUROPEAN VIEWPOINT
Chairman: Riccardo Petrella
Commission of the European Communities, Brussels, Belgium

Telework in the European Community: Problems and Potential
Eberhard Köhler, Rosalyn Moran, and Jean Tansey
European Foundation for the Improvement of Living and Working Conditions, Dublin, Ireland — 221

"Distance Jobs" - A Need for European Action
Werner Wobbe
Commission of the European Communities, FAST Programme, Brussels, Belgium — 239

6. MAJOR THEMES IN THE DISCUSSION OF TELEWORK

Major Themes in the Discussion of Telework
S. Robinson — 245

Telework:
Present Situation and Future Development of a
New Form of Work Organization
W.B. Korte, S. Robinson, and W.J. Steinle (Editors)
© Elsevier Science Publishers B.V. (North-Holland), 1988

INTRODUCTION

Not least because a number of companies and individuals are already experimenting with telework or are utilizing it as their ordinary or entire work organization, the organization of this book moves somewhat unconventionally from the practical to the more theoretical.

It starts with an attempt of Wolfgang J. Steinle (empirica) to raise a few key issues of telework which - according to the value judgement of the author - should be considered when discussing telework.

Part 1 presents the practical experiences of firms with telework - namely F International Ltd. ("Steve" Shirley) with its 1100 data processing specialists in "a company without offices", being the largest of its kind world-wide involved in telework, and the Rank Xerox Networking experiment (Phil Judkins) which is a system of work in which selected and trained volunteers leave their parent company and found their own business, which then contracts to provide specified services to the parent company among other clients, and uses a microcomputer link to do so. These are followed by results of an intensive research on the "Organizational Development of Teleprogramming" from Wolfgang Heilmann, he himself being the initiator of a telework project in his own company INTEGRATA GmbH.

Part 2 moves towards the economic and social factors of telework referring to the corresponding extensive research results and experiences from the United Kingdom from Ursula Huws (empirica, United Kingdom). Margrethe H. Olson, based on her long-lasting research in this area, elaborates on "Organizational Barriers to Telework as an Employee Work Option" putting a special emphasis on organizational and management issues when addressing current trends in telework in the United States. The section is concluded by Marilyn Mehlmann's analysis of "Social Aspects of Teleworking" describing an experimental corner office project, a survey by an employers' association, and an ongoing survey by the federation of white-collar unions.

Part 3 focuses on organizational and technical aspects with Gil E. Gordon examining factors that have accounted for the growth of telework and the role of technology as well as the government in the United States by also presenting some of the likely future scenarios for telework. Barbara Klein refers to the results of a German research project concerned with the evaluation of decentralized work places by means of teletex.

Part 4 offers a look into the future of telework by also giving some insight into the historical background of homework in the Federal Republic of Germany (Herbert Kubicek/Ulrich Fischer) and the current situation of telework. Werner B. Korte (empirica) refers to results of extensive empirical research conducted by empirica mainly for the Commission of the European Communities in the frame of its ESPRIT and FAST programmes and the European Foundation for the Improvement of Living and Working Conditions in Dublin on the resource framework, current situation, with a special emphasis on the inception phase of telework and the likely future situation of telework. Helmut Drüke presents some interim results of a study concerned with the "Decentralization of White Collar Tasks by Means of Modern Information and Communication Technologies" whereas Gerard Blanc elaborates on corresponding experiences from France by also examining the major existing and emerging socio-cultural values in the frame of telework.

Part 5 "Telework from a European Viewpoint" reports on the main findings of telework research projects which were sponsored by the European Foundation for the Improvement of Living and Working Conditions (Eberhard Köhler) and the Commission of the European Communities in its FAST Programme (Werner Wobbe) attempting a tentative developmental forecast and an elaboration of possible European courses of action from the Commission's viewpoint.

Finally, part 6, conducted by Simon Robinson (empirica, Bonn), presents some major themes which did arise from the conference discussion.

It is our hope that the comments and experiences of all the contributors to the conference - which was the first comprehensive international one on telework all over the world and a unique occasion - will foster a more realistic understanding of the implications of telework. We are sure that it will be followed by numerous national and international conferences on this topic addressing special issues of telework.

Welcoming Address

Jan Roukens

Commission of the European Communities,
Brussels

Telework will gradually and increasingly become an element of work organisation, embedded in an ever more important aspect of it: that of communications.

Telework might be described as a technological phenomenon, but in essence it is a socio-cultural phenomenon made possible by advances in information technology and telecommunications.

It could further be described in an objective manner on the basis of observations. This would require proper definitions of what is being discussed. In any case, the conclusion would be that in absolute terms telework does not yet play a significant role in the organisation of labour today and consequently does not have a major impact on socio-cultural relations in the industrialized world.

The telework concept, however, as it is understood intuitively by us, is of an ideological nature and its origins are deeply embedded in our views about work, social behaviour and freedom.

If "work" is a necessary condition for welfare, and this seems generally accepted, then at least we should be free to choose when and where to work. And as certain technological advances appear to provide means to bridge time and space, we should employ these for a more liberal organisation of labour, as long as this does not effect negatively work performance.

If one chooses the ideological approach in discussing telework, rather than the phenomenological approach, it may be made plausible that progress will be slower than one would expect "at first sight".

One reason is, probably surprising, the yet insufficient technology and our inability to utilize technology properly. Some people will claim that they "telework" using very simple local equipment and standard telephone lines. Although nobody will deny that this is adequate, in a number of cases requiring such down-to-earth procedures as text message exchanges, most of the communication functions which characterize our work in groups are not supported by such simple devices and telephone services.

A number of experiments performed in laboratories these days, employing truly multi-media workstations and broadband connections which allow "face-to-face" communication among groups of people, do approximate much closer what would be required in an ideal telework setting.

There will be a day not so far off that this technology emerges from the laboratory environment. But it will take more time until it becomes economically available for a wide public, and much longer to integrate the equipment and these services in another work organisation.

The major delaying factors are to be found in the socio-cultural domain itself, however. In a period of economic stagnation with a rather high level of unemployment, it becomes increasingly evident how important "work" is in our society. Having a job pro-
a certain status, and the work environment itself provides
a playground where people can fulfill ambitions and experience social relations. And for the workers who have difficulty in organising themselves, the environment at the workplace keeps the individuals en route.

To substitute these elements by others of equal social and psychological significance would require an evolutionary period in the order of a century.

Assuming we want that, and that we want it in the context of the telework concept, progress will be steady and relatively slow on the average. The implementation trail will develop just as water finds its way: at each stage, certain directions are more obvious than others.

These thoughts lead us back to the phenomenological approach to telework, and to the majority of the presentations at this conference.

They are interesting as experiments in teleworking, and they show that in some restricted domains, telework may already be considered an obvious alternative.

Welcoming Address

Bernd Kramer

Federal Ministry of Research and Technology, Bonn

Ladies and Gentlemen,

On opening your conference, "Telework: Present Situation and Future Development of a New Form of Work Organization", I should first of all like to take this opportunity to convey greetings and best wishes to you on behalf of the Federal Minister for Research and Technology, Dr Heinz Riesenhuber. He has taken great pleasure in supporting the holding of this conference: not simply by supporting the idea of such an event, but also by providing financial backing. Operating within a spectrum of research work being conducted in the field of technology assessment and system analysis that has been expanding over the last four years, the Federal Ministry for Research and Technology is now about to promote, stimulate or commission scientific studies designed to assess the advantages and disadvantages of new technologies in the context of their widespread application.

If you have followed this research work in the Federal Republic of Germany, you will be aware that the Federal Ministry for Research and Technology - in collaboration with the Federal Ministry of Economics and the Federal Ministry of Labour and Social Affairs - has commissioned a broad-based research programme to look into the effects of new technologies on the labour market. The data already compiled is currently being analysed and is highly representative with respect to technology diffusion and workplace developments in the production and service sectors. The research programme is intended to bring about an important scientific advance in two areas:

- It establishes a continuous observational chain, reaching from the company level (where innovations are made) right up to the macro-economic aggregate level.

- Observed innovatory activities are analytically linked to the resulting changes in the structure of economic performance and activity (including the related shift in demand).

It was in the context of this research programme that we also considered the merits of including the telework sector as a substantive element. However, it very soon became evident that in this field a great deal of empirical and conceptual work was already being carried out. The research projects have progressed to the point where their existing findings and maturing results could only have been fitted into the concept of the aforementioned programme with the greatest of difficulty.

The study on the status of telework in labour law, which was conducted on behalf of the Federal Minister for Labour and Social Affairs, can also be counted among these research projects. In view of this situation it seemed very appropriate to compliment and lend support to the approach taken here. This means first of all ascertaining the state of development by compiling empirical data in this area and, if necessary, organizing appropriate follow-up studies that are better targeted on the basis of a higher level of information.

In this respect, the Ministry for Research and Technology welcomes the fact that this conference was organized on an international basis, in contrast to the studies I have just mentioned. Telework is a form of work organization based around the information and communication technologies and is thereby characterized by the fact that it not only makes it possible to link up a company to a spacially separated workplace or to a complete office at another location in the country, but it also enables trans-border work configurations. This literally means the opening up of new horizons. It is thus hardly surprising that this area has found a permanent place in an international research programme commissioned by the European Community: This place is a component of the FAST research programme, i.e. "Forecasting and Assessment in Science and Technology". Alongside other subjects, this programme has been looking at the field of technology, work and employment since 1984. It is therefore only logical that the European Community is also supporting this conference.

Neither the conference organizers nor its sponsors have the task of judging the issue of introducing telework. First and foremost it is a matter of understanding the various forms assumed by telework, its diffusion and its implications. If chances for enriching the quality of working life or for flexibilizing working conditions can be detected here, it would then be necessary to establish whether government assistance might be needed to realize these chances in a socially acceptable manner. If, having considered all the aspects involved, decision-makers in the economy believe it is opportune to introduce this technology on a wide scale, our task will be to lay down the general framework for social conduct and labour law inasmuch as this cannot already be derived from existing legal structures. This conference represents an important step on the way to understanding the advantages and disadvantages of this technology and the forms of work organization it creates.

I wish you and, indeed, all of us a fruitful working session over the next two days.

TELEWORK: OPENING REMARKS ON AN OPEN DEBATE

Dr. Wolfgang J. Steinle

empirica GmbH
Kaiserstr. 29-31
D-5300 Bonn 1

1.   INTRODUCTION

In contrast with traditional and still prevailing forms of company and work organization, telework makes it possible to bring work to the people or closer to them instead of vice versa.

With the introduction of light and mobile office machinery, the workplace is no longer bound to central office locations. In principle, all types of office work can be executed in decentralized ways either inside or outside organizations. The traditional boundaries of organizations - visible in the form of the physical office buildings which characterize the urban cores of our countries and regions - are no longer relevant [1,2].

On the basis of available information and communication systems, the elusive office is feasible; organizations can live without office buildings. In the early seventies, this technically feasible option led scientists and futurologists to believe that a vast diffusion of electronically based work from the home would occur. In 1971, AT&T came up with the forecast that in 1990 the entire American work force would work from their homes; a decade later, another forecast predicted that in 1990, 50% of the work force would work from their homes [3].

Indeed, it is entirely feasible to decentralize about two thirds of all jobs; in other words, roughly 80 million jobs in Europe could be carried out from the home or from locations near the places where people live [4].

It goes without saying that reality has fallen far short of these aspirations, forecasts or existing potentials. If reality does not conform with theory, something must be wrong with reality. So what is wrong with reality?

A first and major point to be made in this connection is that the technical feasibility of telework does not determine the intensity of practical applications.

Evidently, the uptake and utilization of information and communication systems does not just represent a process of substitution of old machinery by new, but is a phenomenon which depends on such "non-technical" aspects as organizational structures, task and skills profiles and management practices. It is not just a technical domain, it is also closely related to awareness, organizational and managerial culture and other intangibles [5].

Telework is not so much a technical or technological issue but one of organization. The question is, how to organize telework in a way which satisfies social and individual aspirations and requirements, which meshes with societal trends, and also meets economic targets. In fact, telework - depending upon its form of implementation - can be beneficial in social, economic and human terms.

However, in looking at past and present applications, it is quite evident that telework is often not used to satisfy this set of benefits at all but, on the contrary, is implemented within a framework of undesirable working conditions, including lack of social security, isolation of individuals, casual or temporary work arrangements [6,7]. These and related observations have biased the discussion on telework towards issues which hardly relate to technology but to its utilization, which is indeed the crucial aspect. The polarization of positions in this discussion, however, has also led many people to ignore the manifold possibilities and potentials involved in the various telework options. In addition, scientific and political discussions often deal with the symptoms but not the cause.

This is the context in which Empirica organized the conference on which the present reader is based. Its objective was to establish a broad picture of past, present and potential future trends and the possibilities of telework as a new form of work organization.

The present paper does not pretend to summarize the issues discussed at the conference, nor does it endeavour to synthesize a common denominator of the manifold opinions and results. Rather, it tries to raise a few key issues which - according to the value judgement of the author - should be considered when discussing telework.

2. TELEWORK - SCOPE AND RELEVANCE

The potential for telework is vast. As representative surveys in the major European countries illustrate [4,8,9,], roughly two thirds of all jobs lend themselves to decentralization in one form or another. In theory, all jobs and tasks which are not production-related can be carried out in satellite offices, on the basis of electronic homework, neighbourhood centres or other decentralized forms instead of in central office buildings.

Moreover:
- 1/3 of decision makers in European companies are interested in telework options.
- 25-30% of the workforce already working with IT equipment are interested in telework.

Despite this vast potential, however, actual applications of telework are rare. Estimates of numbers of teleworkers in Europe vary between a few thousand and over 100,000. Certainly, the definition, form and scope of the telework under consideration plays an important role with regard to these estimates. Nevertheless, whatever definition one may adopt, it is clear that the existing potential for telework applications has not by any means been reached. As some authors put it: there are more people doing research on telework than there are actual teleworkers.

Against this background, the question is whether telework is an issue worth talking about at all. It is argued here that telework is a relevant concept – however, only if it is seen in a broader perspective and with due consideration to its organizational, social and economic framework.

To some extent, the apparent difficulties in characterizing and defining telework can be attributed to the circumstance that it is not useful to approach the subject from an isolated perspective whether technological, organizational, social or economic. For instance, the most extreme form of telework – electronic homework – does not represent anything new from an organizational or a social perspective. Work from the home has been quite common among a variety of occupations and professions, both among the highly skilled labour force (such as architects, artists, journalists, etc.) and among those with basic skills in low-paid jobs. Seen from the technological angle, telework is not new either; even in its narrow sense it only implies the combined use of electronic information and communication systems.

Therefore, it does not seem appropriate to define telework along any isolated parameter or dimension. The form of work organization, the location of work, contractual arrangements, etc. are all relevant or necessary parameters to define telework; however, they are not sufficient if seen in isolation.

From a pragmatic point of view, the essential question is of what is changing due to telework, and whether the ensuing developments are feasible, likely or desirable. To answer this question, it is essential to consider the framework in which telework is or will become a relevant option.

In looking at telework from a broader perspective, it is essential to be aware of certain global trends:
- mass production has been becoming increasingly possible without mass employment;
- production-related activities are increasingly displaced and substituted for by information related tasks;
- rapid changes in the market demand flexibility, innovation and adaptation.

The continuum of telework options includes forms and practices which are viable and desirable, which are relevant issues from an economic, policy-related or scientific angle. However, it also includes options which are not desirable or which are not viable in the longer-term.

Telework is not an issue for debate because it has emerged or occured "ceteris paribus".

With regard to telework, "ceteris is not paribus". At the least, it is not fruitful to discuss it in such a perspective. Telework is becoming an issue and an option in a specific economic and social situation, and state of technology. For a fruitful discussion of telework, it is essential to understand the mechanisms and workings of the social and economic fabric which lead to a constellation making telework something worth thinking and talking about.

Only if this is achieved will it be possible to delineate those forms of telework which are not desirable or not viable and if so, why, and to define those actual and potential future applications which are viable in economic terms and desirable socially or individually.

3. ORGANIZATIONAL CHANGE AND ITS FRAMEWORK

Even if technology had been at its present state 25 or 30 years ago, it is very unlikely that anybody would have thought or talked about telework. The economy was buoyant and growth was characterized by the presence and expansion of large companies. There was a shortage of labour and unemployment was frictional or hardly significant.

In a situation of expanding markets and rather steady growth, large organizations, strongly hierarchical and tayloristic, are highly efficient and effective. There is no need or economic rationale for decentralization - the more people work at one place the better.

This situation has changed profoundly in recent years: there has been high unemployment, mostly of a structural nature: a sluggish economy, shrinking but rapidly changing markets and other major changes. In this environment, organizations must be flexible and react promptly to changing external conditions to keep pace. At least in terms of employment, the growth of large corporations with centralized and highly hierarchical types of organization has passed. Profit centres, "small is beautiful", intrapreneurship, are some of the concepts which have emerged as organizations have responded to market related changes, changing corporate cultures and management structures, and new economic styles.

These developments are multi-facetted. However, with regard to telework, there is one key component which is relevant: smaller units become vital to organizations trying to become or remain flexible. Large, centrally organized units cannot be flexible, they are efficient in terms of operative or quantitative adaptation. However, they are not efficient in terms of strategy or structure. In a highly dynamic environment, large centrally and hierarchically organized units lack flexibility.

The trend towards smaller units in the economy is not just that companies or organizations are decreasing in size but that their very nature and structure is changing. Smaller units are not mirror pictures of large ones. There are also changes in such things as task profiles, cultural skills, management styles, internal and external flows of information, communications networks.

Generally speaking, smaller units of economic activity presuppose and involve increased interactions and communications both within the organization and with the external environment: The degree of centralization of organizations is not only a key determinant of their structure but it also determines the quality and quantity of information which is communicated between the units - whether individuals or departments - in such a structure. Decentralized organizations use more interactions (quantity) and more complex internal communications (quality) than centralized ones. This also implies job enlargement and job enrichment instead of tayloristic division of labour. As can be seen from the figure below, new information and communication systems are a precondi-

tion and a supporting tool for this tendency. On the one hand, new information and communication systems tend to increase interactions between individuals or departments; on the other hand, increasing interactions and exchange of information are facilitated by the utilization of such systems [10].
It is against this general background that telework is discussed in the present paper. Below, telework in its extreme form of electronically based work from the home is considered. This is followed by a discussion of which, if any, forms of telework seem most likely to be able to facilitate the organizational changes outlined above.

Stage	Communication Media	Organizational Form	Flow of Information	Characterization
1	mainly written communication		↓	strict hierachical, vertical organizational form, fixed/inflexible communication and decision channels; highly specialized and strictly limited functions of the employees; low mobility
2	written and oral communication (telephone), exchange of copies etc.		↔↕	horizontal organization with functional areas; shorter ways for decision with more room for faster decision making
3	see above, but with accelerated communication by means of electronic support		↕↔	network organization; task oriented with flexible rules for decision making; employee qualification not limited to single tasks
4	highly electronically supported communication: terminals, teleconferences; Office information systems		◇	widely spread (also geographically) organization having a complex network very adaptable and innovative; no fixed decision and communication channels

*) Centre for decisions

Source: Rauch, 1982 [10]

FIGURE 1:
Information, Communication and Organizational Structure

4. THE EXTERNALIZATION ISSUE

Externalization mechanisms form a central focus of the scientific and political debate on telework. These can be observed in three areas of concern:

- externalization of labour (freelance or temporary work arrangements for previous employers);
- externalization of work (shifting work from producers to consumers);
- externalization (or increased utilization) of service functions and of producer- related services.

The telework debate is strongly biased towards the issue of externalization of labour - i.e. electronically based work from the home on a freelance or temporary basis. In the present paper, however, externalization of work and of service functions are deemed equally - or even more - important as issues in the future of work.

4.1 Electronic Homework as an Externalization Strategy for Routine Jobs?

In dealing with the externalization of labour issue, it should not be forgotten that it concerns one particular form of telework and not telework as a whole.

The most extreme form of telework - electronic work from the home - is a matter of controversy in both scientific and political debate. Critical voices argue in favour of a complete ban; in some circles, "telework" has, with some basis, become a dirty word and a negative symbol of technical progress [6].

On closer investigation, it becomes apparent that this negative image has arisen not so much because the place of work is an issue of the ongoing debate but rather because of the contractual or work-related and personal arrangements which prevail, the type of work carried out, and the level of pay.

Telework is often automatically assumed to mean freelance or temporary work from the home, monotonous tasks, low pay and increased stress through the attempt to combine work with household responsibilities [7]. A housewife with child-care responsibilities, entering data or text on piece-rates, working for her previous employer (where she had a contract as an employee) is the prototypical negative symbol of a teleworker.

As a matter of work ethics and social responsibility, this type of telework is not desirable at all.

This extreme type of telework, however, is a symptom, not only of bad employment practice, but also of managerial lack of flexibility and inertia.

Telework in the form outlined above may be an intermediate short-term solution, but not a successful longer-term option. It is, and will remain, a marginal form of work for several reasons:

- it only lends itself to highly standardized routine tasks;
- it perpetuates existing organizational and managerial structures instead of adapting them to changing economic, social and individual requirements;
- the tasks and work contents concerned are increasingly automatized (data and text entry, and, in the longer term, programming).

As has been seen in section 2 of this paper, new information and communication systems facilitate and reinforce the trend towards flexible smaller units, they ease communications, both internal and external to organizations, and they support the move from vertical to horizontal hierarchies, and from routine, segmented tasks to complex multi-facetted tasks. These potentials, which are essential in the shaping of competitive units of economic activity in a climate which requires rapid adaptation and flexibility, do not depend on the externalization of labour.

4.2 The Off-Shore Threat

Like the trends we observed during the past in declining and labour intensive sectors, telework is likely to constitute an activity which has to be seen in a worldwide perspective. The trend observed during the past has been one of regional decentralization of low skill, low-paid activities to peripheral areas in the advanced economies. The next stage has been the externalization of such work to developing countries, followed by a third stage of reintegration using sophisticated methods of production in the advanced economies and their environment. For instance, with regard to textile industries, the past trend was one of externalization of production units to depressed regions, followed by a second wave of externalization of labour intensive activities to developing countries and a third stage of reintegration of high quality production back in the advanced economies. An indicator of this trend with regard to telework are off-shore activities such as data entry and text processing in the Caribbean and Asiatic regions.

Currently, there are at least 12 U.S. companies with data processing operations in the Caribbean. The largest such off-shore operation employs about 1200 people (in Haiti). Other examples include operations in India, China, Singapore, the Phillipines and Mexico. The major motive to go to off-shore relates to labour costs: data or text entry costs are 75-90% below European or US American standards in the above mentioned regions [11].

Governments in these regions are quite aware of the fact that this market is temporary and will decrease in size with increased automation. They expect, however, that off-shore activities in their countries, if coupled with appropriate training and skills-upgrading policies, will allow their labour force to adapt to the pace of technical change and thus open up new opportunities in the future.

4.3 Externalization of Work - Electronic Homework and the "Self-Service Economy"

Telework has profound implications for the very notion of work. On the one hand, it tends to wipe out the traditional demarcations between work and leisure. On the other hand, it tends to shift work from producers to consumers.

The surge of do-it-yourself activities has brought about a redistribution of work from producers to consumers. Similarly, telework tends to shift work from the producers of services to the consumers. The so-called "self-service economy" of which this tendency is an instance, is due both to potential savings of labour costs on the producer side and increased availability of time being spent on self-production or procurement of services on the consumer side.

With regard to telework - especially in its extreme form of electronically based work from the home - this type of externalization, at least from the point of view of time budgeting, does not so much concern work content as leisure: so-called "obligated time" is creeping up at the expense of leisure; for instance in the form of home banking, electronic booking, shopping etc. [12].

Toffler goes even further in suggesting that, in the future, consumers will become "prosumers" in producing the goods they want via electronic networks linked with computer integrated manufacturing facilities.

4.4 Polarization of Labour Markets – Externalization of Labour Versus Externalization of Service Functions

Electronically based work from the home is an issue in the polarization of labour markets. To put it bluntly, this extreme form of telework is or may be advantageous for the highly skilled primary segments of the labour force and is and may be detrimental to the secondary labour market: it establishes barriers to entry through isolation and it isolates the less-skilled, thus marginalizing this segment of the labour force [13].

Electronic work from the home as an externalization strategy segregates the work force into those with and those without personal, informal contacts with other employees. More recent advances in organizational theory, by contrast, demonstrate the relevance and decisive role of informal communication in and for organizations: people learn more about "what is going on" during lunch time or coffee break than through "official" top-down information channels.

Telework from the home, if practised as a strategy for the externalization of labour, does not allow employees or ex-employees working freelance to adapt to the need for increased complexity of work or job enrichment, which is essential for the future. Instead, it perpetuates existing organizational mal-adaptation.

Telework plays a substantial part in the process of polarization or segmentation of labour markets. At one extreme, housewives in low-income households with child-care responsiblities represent a negative "ideal type" of telework for low-income and unskilled labour. At the other, those with scarce skills in high-income population groups, like data-processing professionals can use telework to increase their options and enhance their labour market status.

Externalization need not necessarily be negative, however. It can also create new opportunities. These relate to the emergence and shaping of new activities.

Telework is not just an issue which is relevant with regard to the restructuring of existing organizations; it also creates the potential for new activities to emerge by exploiting the combined applications of information and communication systems, and utilizing the competitive advantages which such options may constitute. In this case, traditional markets may be entered on the basis of new products and services and a new form of work organization. The case studies carried out at Empirica illustrate that a substantial proportion of telework implementation has not been realized in existing companies or units of economic activity but is the result of individuals deciding to use this technology for gainful employment and business creation [13].

For instance, data processing professionals and engineers belong to segments of the labour force where demand clearly exceeds supply. In the Federal Republic of Germany, Siemens alone (one of the largest employers in the field of information and communication systems) needs to hire 40,000 programmers. This situation increases the pay-levels and improves the labour market status of those concerned.

This situation greatly facilitates the creation of new businesses. Indeed, an analysis of existing telework applications indicates that roughly 4 out of 10 cases of employment in this area of combined information and communication technology are in new businesses, established to deliver producer-related services [4]. Hence this type of externalization - in contrast with externalization of labour strategies - may produce positive results in terms of job creation. It also meets the requirements of smaller companies for whom it is not feasible to establish such services in-house.

5. TELEWORK AND ORGANIZATIONAL CULTURE

In discussing the various facets of externalization, it has become apparent that telework plays an ambiguous role. The externalization of labour does not seem to be a viable longer-term strategy, but the externalization of service functions may initiate new economic activities.

The question remains to what extent telework is an option worth considering within existing organizations by decision makers or managers and their staff. To put it another way: to what extent is telework an option for existing organizational units not adopting externalization strategies of one type or another?

The traditional company organization, currently by far the most dominant in the Western World, is characterized by a hierarchical coordination of tasks that integrate workers in the hierarchy of an organization by employing them and paying them fixed salaries [14].

Olson describes the traditional organization as work places where it is "assumed that a critical mass of employees will occupy a central work place a set number of hours a day, typically "nine to five". Work performance and organizational procedures are critically bound by this place and these hours" [1].

In a decentralized complex work environment, management - virtually by definition - cannot control individual work steps, it has to be output-oriented, it must be able to delegate responsibilities and it must oversee a complex network of interactions. Organizations with strict vertical hierarchies and control-oriented management are clearly very unlikely to meet these requirements.

As an analysis of existing telework applications demonstrates, this form of work organization evolves both in highly centralized organizations and in highly decentralized ones. However, the scope and nature of the telework carried out within these two extreme types of organizational structure is of a distinctly different nature. In highly centralized structures, due to the organization of work and tasks by isolated work steps without an overview of a single responsibility for an entire process, the contents and form of telework remain narrow and limited in scope. Accordingly, the type and complexity of tasks differ substantially.

5.1 Organizational Aspects of Hard- and Software

Technical and technological development lags far behind the user requirements. Hardware and software design is generally based on the practices of large corporations and the type of centralized work environment which they foster. On the one hand, this may involve the risk of systems design being oriented towards the functional requirements of such organizations while ignoring the ones of decentralized user environments. On the other hand, the movement towards smaller organizational units is quite recent. Similarly, the design and implementation of technical systems for such units is at an early stage, too.

The micro-computer was only introduced for work-related purposes in the early eighties; networking equipment for micro-computers became available late in 1986; electronic communications networks and services are still at an early stage of development; ISDN which allows for simultaneous transmission of data, text, voice and (real-time) imagery will only be operational in 1988 in the more advanced economies. Hence, the type of equipment suitable for work organization in a decentralized office environment is only at the stage of initial implementation.

5.2 Management Inertia

The role of managerial and corporate culture cannot be ignored in discussing the issue of telework. Managers traditionally measure their status by the number of employees surrounding them. Under the most extreme form of telework - electronic work from the home - traditional presence (physical presence of employees in usual office environments) - is, as a matter of fact, no longer feasible [15,16].

It is a status symbol to have subordinate employees around and to show this situation to visitors: the more there are the better. From this point of view, employees working remotely do not count. In this regard, not only do management attitudes need to change but also management styles and skills [8].

A reasonable strategy to implement telework presupposes a variety of organizational and managerial changes and amendments.

If new forms of work organization are introduced without adaptation of management, the result is far from optimal. New possibilities like treating business correspondence via linked terminals do not only involve a new job profile for secretaries or typists. Where a secretary no longer submits mail to her boss but deals with it herself at a terminal, certain delegation processes cease to be necessary, competences and decision functions must be reallocated. Quite often, this does not happen because management is not aware of its changing role and the need for adjustment.

5.3 Decentralization of Production and Concentration of Control?

In the discussion about the impact of telework on company and work organization, the issue of decentralization occupies a central role. Very often, it is predicted that economic activities will become geographically scattered and that companies will decentralize their operations for reasons of client proximity and reduction of transport costs.

Indeed, as already mentioned, telework makes it possible to bring work to the people and not, as has been traditionally the case, people to the work. However, the assumption of increased scattering of economic activities, suggested, for instance, by Toffler, as a major future trend, to some extent ignores the fine line between decentralization and deconcentration. The geographical dispersal of activities which is involved in deconcentration and the dispersal of hierarchical functions which corresponds to decentralization are quite evidently two distinct phenomena.

Trends towards the deconcentration of production functions - in a large sense comprising products, services or information - can indeed be observed. However, with regard to control or decision functions, the patterns which have become apparent are different from those observed in the area of production.

In general, one can observe a trend towards the deconcentration of production-related activities coupled with the centralization of control and decision functions. This tendency can certainly alleviate problems of unequal geographical distribution of economic activities, however, it cannot change the geographical distribution of functional dependencies. In fact, the reverse might very well be the result.

For instance, because of improved telecommunications services and networks, managers of branch plants may incur increased control from the parent company rather than an improved flow of information [17].

This, however, can also be seen as an issue in management inertia: if supervision results in increased control, the raison d'etre of decentralization - which is flexibility and improved adjustment to rapidly changing markets - is no longer present. Again, this way of proceeding perpetuates existing mal-adaptation instead of overcoming it by more appropriate arrangements.

6. CONCLUDING REMARKS

Telework is a more extreme form of decentralization than the establishment of profit centres, intrapreneurship, or other new organizational forms. Thus, in discussing telework, it should be kept in mind that we are dealing with an issue which, in the phases of introduction of new information and communication systems, presupposes an advanced state of IT uptake and diffusion. In a time sequence telework follows developments in other areas of IT application and of organizational change.

This also implies that the telework existing and observed today must not be the telework of tomorrow. Although scientists and forecasters often conclude that the observed developments will continue in the future, this type of proceeding certainly lacks imagination, and in the light of ongoing, global and individual developments is not too reasonable.

The telework of today will not be the telework of tomorrow either because we do not want it, because some activities may only be relevant in the short-term market, or because of a more complex phenomenon relating to changing organizational and managerial culture.

It has been argued in the present paper that telework in its extreme form of electronic home work is not a viable longer-term option, at the least not if seen or used as an externalization strategy with regard to employees previously working at central office locations (externalization of labour). Only monotonous, simple or routine jobs lend themselves to this type of externalization. It is a short-sighted response to a changing economic environment which asks for long-term adaptation.

Long-term adaptation requires changes in management attitudes, company organization and economic style.

There is a clear need for more flexible units of economic activity. This involves decentralization, less hierarchical and more communications-oriented work and company organization. Telework may support and facilitate these developments.

The sluggish diffusion of telework can be referred back both to the circumstance that applications are often used to perpetuate existing structures instead of adapting them, and to management inertia, which tends to maintain existing forms of company and work organization.

Telework can only become a viable reality if management inertia is transformed into more active adjustment.

REFERENCES

[1]   Olson, M.H., 1982: New Information Technology and Organizational Culture. In: MIS Quarterly Special Issue, pp. 71-92

[2]   Gordon, G.E., 1985: Telecommuting: A Managemenet Challenge. In: Data Processing Management. August/September, pp. 1-11

[3]   Karcher, H.B., 1984: Büro der Zukunft. Dominanz der Mikros und der Multifunktions-Endgeräte. In: Office Management Nr. 10, pp. 882-887

[4]   empirica, 1987: Profiles of the Population Potentially Concerned with Telework - The Supply of Teleworkers. Results of the Employed People Survey (EPS). ESPRIT project 1030, empirica working paper no. 6, Bonn

[5]   Papers by M. Mehlmann, U. Huws and M.H. Olson in this publication.

[6]   empirica, 1986: Telearbeit - Meinungen und Standpunkte der Sozialpartner und der Erwerbstätigen sowie das Potential dezentraler informationstechnisch gestützter Büroarbeit in Europa. Final Report for the European Foundation for the Improvement of Living and Working Conditions. Bonn

[7] Huws, U., 1984: The New Homeworkers. New Technology and the Changing Location of White-Collar Work. Low Pay Unit Pamphlet No. 28

[8] empirica, 1987: Market Potential for Dezentralized Office Work Based in Information Technology - The Demand of Telework. Results of the Decision Maker Survey (DMS). ESPRIT project 1030, empirica working paper No. 7, Bonn

[9] empirica, 1987: Die Ausstattung europäischer Betriebe mit Informations- und Kommunikationssystemen (COMTEC). ESPRIT project 1030, empirica Working Paper No. 8, Bonn

[10] Rauch, W.D., 1982: Büro-Informations-Systeme. Sozialwissenschaftliche Aspekte der Büro-Automatisierung durch Informations-Systeme. Wien, Köln, Graz 1982, p. 76

[11] US Congress, Office of Technology Assessment, 1985: Automation of America's Offices., pp. 211f. Washington, D.C.

[12] Rajan, A. 1987: Services - The Second Industrial Revolution, London, pp. 56f

[13] empirica: A Survey of Teleworkers and Teleworking Companies - Case Studies in Current Telework Environments. empirica Working Paper No 11, Bonn 1987

[14] Brandt, S., 1983: Working-at-Home: How to Cope with Spatial Design Possibilities caused by the New Communication Media. In: Office: Technology and People, 2, pp. 1-13

[15] Judkins, Ph. et al. 1986: Networking in Organizations. The Rank Xerox Experiment

[16] Olson, M.H./Lucas, H.C., 1982: The Impact of Office Automation on the Organization: Some Implications for Research and Practise. In: Communications of the ACM, November, Vol. 25, No. 11, pp. 838-847

[17] Kubicek, H., 1985: Die sogenannte Informationsgesellschaft. Neue Informations- und Kommunikationstechniken als Instrument konservativer Gesellschaftsveränderung. In: Arbeit 2000 - Szenarien über die Zukunft der Arbeitsgesellschaft. Hamburg

1
PRACTICAL EXPERIENCE

TELEWORK IN THE UK

Mrs "Steve" Shirley OBE, B.Sc., CBIM., FBCS., FRSA

F International Group plc
The Bury, Church Street
Chesham, Buckinghamshire
HP5 1HW, England

Telework fits squarely into the social and economic trends shaped by the convergence of computing and telecommunications. This paper concentrates on British experience and is based on the author's 25 years within the unique F International systems and software group.

1.   INTRODUCTION

The author has dual experience of telework or work at a distance. She founded F International in 1962 as a teleworking organisation which has prospered to become the world's most successful model of this way of working, and she herself has worked from home for nearly a quarter of a century, performing a variety of tasks from technics to strategic management.

The very first example of telework was a company called Computations Inc. which as long ago as 1957 offered the scientific community around Boston the services of a small group of home-based computer experts. It prospered for many years under its founder Elsie Shutt but apparently did not endure longterm as researches can find no trace of it today.

But it is surely significant that the first steps towards telework were taken within the computing industry. Not only are computer people closely involved with the technologies that encourage telework, but many computing jobs are well suited to the practice. The industry is growing rapidly, thinks internationally and is full of people young in years and spirit.

Yet a discussion of telework is not about computer technology as such. Computers are becoming merely the nodes of the network of telecommunications and it is networking which is the fundamental technology supporting the spread of federated workstyles like telework.

Further, telework is but one small part of a widespread trend towards many new styles of working.

2.   THE CENTIPETAL CHANGES IN WORK

Everything about work is changing: how we work, when and where we work, why we work and how it is compensated. There are also changes in who is doing the working and basic rethinking about the real nature of work.

To concentrate on the place of work, a vast bulk of people used to work within large corporations, in state-owned enterprises and in the public sector. Today, there is a swing towards the small business sector, size coming right down to the individual free-lancer or entrepreneur. People increasingly work for service organisations rather than in manufacturing and agriculture.

And 'cottage industry' once written off as a source of work - the word is "work" rather than "employment" because the contractual relationships are also changing - is making a comeback. Other variants of work at a distance include not only branch offices and decentralisation and neighbourhood work centres where people from a number of different organisations congregate on a regular basis but also workshops and business centres providing a range of facilities on short-term hire.

Rather than shift the workplace to different locations, there is yet another variation:- workstations literally on the move.

Since last November, ASEA employees living in Stockholm can work en route to their head office 120 km away. The company owns a custom built carriage, the first private passenger coach on Swedish railways. It seats some 40 people and is equipped with about 20 telephones, typewriters, personal computers and telefax. Staff using this office-on-rails are paid half their normal hourly rates plus financial support equivalent to one-third of the 2nd class fare. Management view it as a recruitment aid and comment that it cuts out the risks of driving to work on Sweden's dark, icy roads. They predict 100 such coaches running on behalf of various companies in ten years.

Bleepers allow occasional tasks to be provided by people on call around the clock. The mobilisation of telephones and their newer facilities, some of which make them indistinguishable from a networked terminal of significant power, are revolutionising the siting of many small businesses and specialist services. The plumber and the vet join the US President in taking their telephones with them.

Functions as well as people can be de-centralised. For example, a sensor linked to a micro mounted on the petrol tanker can generate the printed invoice at the time of delivery of the fuel.

There is a similar flexibility of concept when the timing of people's work is considered. Just as the capital intensive industries attracted people into the cities with their factories and shift work, the labour intensive service industries allow people to return to the country using an infrastructure of telecommunications rather than the transport system of the past.

There are parallels with the land enclosure which took place in the 18th century. This deprived rural people of their access to common land for grazing animals, collecting wood etc. At that time Britain and several other European countries provided some compensation through the garden allotment scheme. Industrial allotments could provide workshop or other facilities so that people could exercise valid skills which our changing society can no longer gainfully employ.

3. THE BRITISH SCENE

Empirica's Ursula Huws [ibid.] is summarising the existing statistics on the trends towards telework in the UK.

The national characteristics which have encouraged the growth of telework since the early 1960s include:

* a densely populated small country (as also Japan)

* yet with large, uninhabited, barren areas
* centred around London in the SE corner

* with an increasing division of culture between North and South

    and

* a tradition of innovation and invention (though not of commercial exploitation

In Scotland there have been government initiatives to encourage distributed office work - both as business centres and via teleworking - in the beautiful but empty Highlands and the Islands in the northernmost part of the country.

4. REGULATIONS & COMMERCIAL PRACTICES IN THE UK

There have been many unsuccessful attempts to persuade unemployed people in the North to relocate to the more buoyant South East. In taking work to the people, grants to encourage the creation of new jobs in areas of high unemployment do not yet apply to home-based employment.

The MSC, Britain's vast central government training department, does however recognise home-working as a viable option.

Local authorities generally turn a blind eye to infringement of the zoning regulations governing the commercial use of residential areas. The trend is for the legislation to be updated in terms of the laws of nuisance. Thus anything which is a nuisance to neighbours will become disallowed - whether in terms of noise, traffic or other hazards. This would obviate the difficulty the author had in the 1970s when ruled to be in breach of planning permission. This forced certain cosmetic changes such as moving a registered office to a 'commercial' location.

The existing UK law is generally confused over the status of homeworkers.

New standards of taxation and the rating situation have to be considered alongside zoning. New Health and Safety legislation will be required. Unlike the Netherlands, no trade union has shown any interest in recruiting teleworkers.

With only 24% of Britain's houses owned outright, most people have to study the small print in the mortgage (35% of houses) or lease with private (12%) or Council (29%) landlord to make sure homework is not disallowed. Negotiating an exceptional clause in a lease or mortgage can be a timeconsuming skirmish with bureaucracy.

Insurance is not an insuperable problem. Certain classes of workers, notably the professions but including also machinists who make up garments, childminders, landladies and a thousand other classifications, have always worked from home. Equipment is insured according to who owns it. Home and third party insurance normally has to be modified since most insurers normally limit their liabilities by excluding paid work or workers at home. Practices may change as Britain becomes more litigious.

Unlike Japan there are no British organisations which maintain or clean the office premises of someone working at home. The cost of equipment maintenance, as of consumables, is normally met by whoever owns the equipment.

5. TELEWORK IN LONDON

With London battling to remain one of the world's leading financial centres, the City is overspilling its boundaries into Dockland. The cost of space is phenomenally high and the recent 'Big Bang' of deregulation has rocketed many fees and salaries.

Many high-profile financial analysts and brokers cut down on travelling time by having the latest market figures fed to terminals in their homes. These are typical 20th century cottagers beside their electronic hearth in their intelligent homes!

The Prudential Assurance Company supplies terminals at home for some of their software maintenance teams so that they can provide 'hot-line' diagnostic support around the clock.

British Telecom is piloting Network Nine business centre services; there are also scores of other examples of work at a distance but none in professional management, each being experimental or involving only a few people. Unsuccessful experiments like Lambeth Council have been discontinued.

6. THE MAIN GROUPS OF TELEWORKERS

There are only three organisations who use homeworkers in any number: ICL, Rank Xerox and F International. These three are co-sponsoring a study of the management of teleworking planned for publication by John Wiley this year. There is also a group of disabled homeworkers who were sponsored by government and who work for a number of different employers.

ICL

This group now comprises over 150 employees, each supplied with a personal computer and is part of the computer manufacturer ICL which began 'off-site' working 18 years ago with its Contract Programming Services Unit.

With skilled technicians in short supply, ICL sought to maximise the skills of its workforce by employing staff to work flexible hours from home. Each is well equipped with both hardware and software and enjoys nearly all the standard ICL employee benefits. This arrangement is well suited to people with family commitments and the workers are almost all women with young children.

Instead of waving goodbye to valuable resources, ICL created a successful and self-sufficient contract software house. Systems support and technical writing functions have also been farmed out.

Although not in profit centres, the homeworkers have survived several corporate re-organisations and can be assumed to be cost effective.

RANK XEROX

The Rank Xerox experience is described by Phil Judkins [ibid.]

F INTERNATIONAL

F International trades primarily in Northern Europe and is a computer systems and software company. It operates as a wholly teleworking establishment and its chartered Mission:

> "to develop, through modern telecommunications, the unutilised intellectual energy of individuals and groups unable to work in a conventional environment"

has been implemented throughout the company.

The company has teleworked through the various stages of the organisation's development, from entrepreneurial mode through semi-autonomous regions on to its current maturity as a managed organisation. Similarly, it has teleworked through various phases of the technology: from the simple telephone through the on-line stage to the current era where the technology and its decreasing cost allows homeworkers to be increasingly equipped with electronic mail and similar aids.

F International therefore offers quite a degree of generality! And one which can be measured in commercial terms. With an annual turnover approaching £10m in 1986, it is ranked 12 in the UK league of software houses. To put this into perspective, the information technology industry is very fragmented, but represents a £6bn turnover and employs a quarter of a million people. F International has 1100 people of whom 250+ are employed and 800+ are freelance (hence the F in the name).

Because the business is computing and telecommunications, the majority of the workers are in these specialisms. But because it is wholly a teleworking company, there are people operating in this way for all the usual administrative and support functions as well as homebased managers and secretaries in both line and staff positions.

Teleworkers are engaged as employees, mostly part-time, or as freelance members of a 'panel' contracted for flexible hours of work on specific projects as and when these arise. The panel averages 75% utilisation of its available hours.

The aim is for flexibility and (in the UK) the company is moving towards annual hour contracts. There are many transfers between the two contractual forms of employee and panel and this reflects individuals changing needs since there is no difference in status. The organisational culture generally separates status from position.

Originally, homeworkers managed with surprisingly little equipment, perhaps just a telephone; now they have at their disposal all manner of sophisticated interlinking office technology, including personal micros, word processors, workstations, printers, fax machines, electronic mail and teletex. In general a combination of traditional systems (telephone, telex and postal and courier services) and higher tech. is used.

F International was one of the first users of fax and is currently implementing what will be one of the largest wide area networks within a single British organisation. This is the recently introduced E-mail which the author has on her desk and her secretary has on hers, as do 120 others with more people being equipped each month.

Contrary to most expectations however, the technology is the least pressing concern. It is the human interfaces which are critical.

Overheads are spent on management and communications, not rent and rates, and as such are variable, not fixed, costs. But F International came to the overheads problem from a new angle: how to access a skilled workforce in short supply, rather than how to get rid of the need for premises.

After 25 years there is still very little in the way of accommodation. The operational headquarters is a simple two-storey building, a converted church hall, with space for a few office based staff, essential equipment and multi-purpose meeting rooms for visiting workers.

Teleworkers provide business flexibility - a key to economic survival today. People are available as and when needed and generally paid in terms of productive output rather than presence. Those covering essential functions, either for F International or its customers, can work 'unsocial hours' if it suits them. Others have a portfolio of activities perhaps doing seasonal work, rearing children, studying or in a variety of ways finding a lifestyle which suits their family. A few are disabled. As described in the following section, one man in a wheelchair was sponsored by the government. Staff wastage is considered to be low for the industry because the workplace is immaterial and people are not lost to the company when a partner has to move location, nor their time lost due to immobilising illnesses.

Some of these advantages are offset by expenditure elsewhere. There are extra equipment costs, communications, and on new forms of training for teleworkers and their managers who must both learn fresh skills. Even where freelancers provide their own equipment, the company may bulk purchase this and allow the teleworker to pay off its cost over a number of months.

The management is simple, tightly structured and well administered.

Is the smartness of the administration an overcompensation for an innate clumsiness in such a dispersed organisation? Or is it perhaps a characteristic of the predominantly female workforce?

Pre-timed and costed project 'chunks', regularly monitored by worker and manager with final quality audits by an independent team, keep teleworking viable despite the growth rate (which has the organisation doubling in size every few years).

The high quality output is demonstrated by the much coveted NATO Quality Assurance registration which was achieved at the first attempt.

Training is ongoing and follows the assessment procedures which are open to all. A variety of training methods are used: some computer based, some open or distance learning, some embedded into ongoing activities. People are also deliberately brought together for induction or training in management or technical techniques; this helps bonding and allows the use of outside consultants to help smooth the process of change.

In a growing organisation, even when tasks per se do not change, individual jobs are changing rapidly; in a changing world, the organisation itself changes - currently towards a less production and more market oriented stance.

There is a balance between the tightness of control and the flexibility of the work methods employed. The individual sees more of the flexibility. Within the obvious constraint of getting tasks done on schedule, people can start working before the commuter trains even start running or when offices are long since closed. Apart from other work and family responsibilities, many F International teleworkers find that they can pursue hobbies and generally tailor their total working lives according to variable interests, thus enriching their lives.

The company has been extensively studied for its unique material and an F International case study is to be taught at Harvard in April 1987.

Many surveys of the homeworkers have been carried out. Over half have children under school age. They do not generally look to the company for social stimulus. There is, however, an active social programme alongside the work activities, including twice-yearly "Freespeaks" in each functional unit when teams are invited to meet with key managers to discuss a range of business topics. Regular local and corporate Newsletters keep people in touch. A Company Day for the entire workforce is set for a Saturday next June, the last such event being four years ago.

F International workers rely heavily on teamwork, even if this means a team as small as two, to overcome the daily, professional isolation. Contrary to expectation, most of the teleworkers do not spend all their time at home. They tend to work from, rather than at, home and also spend time at one of the 11 small offices (one being little more than a business centre) or visiting customers.

One in four of the workforce took the opportunity to acquire shares in the company when F International re-registered as a public limited company in 1985. 20% of the shares are thus held either individually by the workforce, (including those who are office based) or collectively via a trust.

In summary, F International teleworkers are a 'type' able to adapt and thrive on independence and the challenge of working remotely. This would not suit everyone. From the steady stream of applicants who seek such work, F International selects those most likely to keep it a leader in the growing software industry.

7. DTI'S SCHEME FOR DISABLED HOMEWORKERS

The fourth group of homeworkers in Britain was started as a government initiative following the conjunction of the UN's Year of the Disabled in 1981 with IT82 - Britain's Information Technology Year.

This comprised initially six, then twelve now 100 physically disabled people, carefully selected and matched with real vacancies, who were sponsored by the Department of Trade and Industry with sufficient equipment to allow them to undertake that job from home.

The most highly paid was an oil expert working in Scotland for Shell Headquarters in London. Another administers the British Computer Society's systems analysis examinations from her seaside home; she enjoys the job but has commented on its lack of career progression. A third is F International's Bob Campbell, equipped with £20,000 worth of electronic (micro, telephone answering equipment, modem, etc.) who works part-time in its technology division and teaches colleagues and client staff to use micros mainly for wordprocessing applications. The jobs differ widely but are all graduate-status.

8. CONCLUSION

The original reason to separate home from workplace no longer exists. The common law principle of "quiet enjoyment" needs to be re-enthroned. New opportunities abound following the recent convergence of computing and telecommunications. More Britons seek to work at, or near, the home than ever before. It is immaterial that many of them would prefer to be elsewhere or in a conventional work situation.

Despite 4 major examples of homeworking in Britain, there is a lingering prejudice about working at home. In class conscious Britain it is generally seen either as a Utopian option, leading to a re-populated Arcadia with communities of beautiful people, or as a way of making unemployment tolerable.

The ambivalence of homeworking will keep it a peripheral activity until such time as Britain changes its ideas about modern living.

9. REFERENCES

1. Harvard Case Study - published for Harvard Business School 1986.
    References:- F International (A), 9-486-118;
                F International (B), 9-486-119;
                F International (C), 9-486-120;

2. Home. A Place for Work?
   Richmond Postgate
   Calouste Gulbenkian Foundation London 1984
   ISBN 0 903319 26 B
3. Changing Work Patterns
   How companies achieve flexibility to meet new ideas
   National Economic Development Office, London 1986
   ISBN 07292 0788 9
4. The Distributed Office
   Presentation to the Royal Society of Arts February 1987
   Mrs 'Steve' Shirley
   In preparation

Copyright:
F International Group plc
January 1987

TOWARDS NEW PATTERNS OF WORK

P E Judkins MA MSc MIPM

Manager, Great Britain Personnel & Management Services

Rank Xerox Limited

Networking is an experimental system of work whereby selected and trained volunteers leave the parent company, and establish their own limited company, which in turn contracts to provide services to the parent company among other customers, using a microcomputer link as an information processing and communications tool. It is only one of a series of Rank Xerox experiments with new patterns of work; the total series covers networking, intrapreneurship (entrepreneur "buying out" the right to provide services to a company), Xanadu (a network of ex-employee small business suppliers), distance learning systems, a variety of training and development programmes for young people and support staff, an education liaison programme (Alpha-2000) and the application of artificial intelligence workstations to aid junior staff career development.

The networking project itself is small scale (54 people, or 5% of Rank Xerox central staff) and increases steadily at the rate of one or two people per month, the growth rate being dependent on the need for careful selection and training; it is a programme applicable to all ages (current networkers range from 28 to 62 years of age) and all skills (not merely computing : current networkers provide a full range of skills from marketing, market research/planning, business planning, financial work, major capital programme control, tax advice, recruitment, safety, security, pensions to public relations and corporate affairs). Most networkers are managerial, professional and executive levels - that is, predominantly information workers, for remote working does not support manufacturing or flow-line work. The system is very young (only three years old, with most participants more recent) but has already developed many variants; it is not, in any sense, a single, invariable "solution" to future patterns of work, for these will contain as many variations as the "traditional" contract of employment (itself widespread for only some two hundred years).

There were three origins of networking - the business need to cut overhead costs to survive and grow; the requirement to meet the needs of people to regulate their own work; and the development of technology, making telecommunicating microcomputers available cheaply.

Examining each of these in turn, we found in 1982 that our facilities costs (rent, property taxes, energy costs, maintenance, security, depreciation etc) absorbed one-third of our Central London Headquarters spend, with a further third on salaries, one sixth on employment costs (taxes, pensions and benefits) and one quarter on all other costs. In individual terms, an individual with a salary of £10k probably saw a net pay of £6,700; but the company saw a cost, fully loaded with facilities, taxes etc of £27,000. Not only was the scale of these facilities costs significant - they were inflationary, rising faster than CPI; they were, in the main, dependent on factors outside company control, such as political decisions. The spend did not, within limits, add much to staff motivation; and the enormous sum expended on these facilities was sterile, for it added nothing to the company's product in the marketplace. A solution had to be found - but how?

In seeking the answer, we were fortunate first of all in our workforce and our organisation, both of which are youthful with a highly entrepreneurial perspective which encourages novel approaches to problems. Within such an organisation, many people would consider it a logical part of their own development to found their own business, and many have in fact done so, displaying an enhanced productivity and high motivation when they did (and they were not unproductive or demotivated before!) However, when we attempted to retain the scare skills of some of these ex-employee entrepreneurs, (for it was often those people with a creative high level skill who moved on) and used the traditional contract of employment to do so on a part-time basis, the experiment did not work; the individuals were unsure of goals, and impatient of what they saw as restrictions on their freedom, and their managers unsure of how to manage, motivate, and meld into the work group individuals who appeared only on a part-time basis. Part of the answer, therefore, seemed to be to capitalise on this entrepreneurial drive and let people regulate their own work; but the traditional contract of employment did not seem to be part of the solution.

Finally, communication was an issue which had to be solved - not technologically, because remote terminals to a mainframe computer allowed remote working some twenty years ago, and a number of firms practice this; but cheaply, because the need for an expensive mainframe or timesharing agreement would obviously reduce the facilities savings. Fortunately, in 1982, the answer was at hand - the telecommunicating micro had just become cheaply available, and its costs even then looked set to reduce further in the same way as the pocket calculator price had slumped to the level of disposable technology in a single decade.

With these elements of the answer to hand, we began by dividing the jobs of central staff into two categories.

- those in 'continuity' mode, where being physically present at a specific place was essential to the task being fulfilled; a front entrance, receptionist, or cash office counter clerk, are examples.

- those in 'output' mode, where the organisation was interested in the output of a person's job, and the place where it was performed was less relevant. A computer programmer is one practical example : a company needs a program which performs certain tasks, and where the programmer sits when creating it is less relevant.

Networking, therefore, sought to cut out the large and sterile facilities costs for the output mode jobs, where facilities were less relevant; and we sought a means whereby output mode workers would work from a remote location (e.g. home) via a computer link.

The means employed built on the entrepreneurial characteristics of our workforce, and our bad experience with the traditional contract of employment. The rules we laid down specified that the individuals interested in working remotely would leave Rank Xerox employ, and establish their own Limited Companies. The ownership of this company gave them an incentive to grow with the company and maintained the enhancement and productivity gains in their working for themselves; by contrast, simple homeworking achieves a saving for the parent company, but contributes little to the homeworker, and self-employment does not generate the same motivation to build a bigger organisation. This networking company then contracts with Rank Xerox to deliver specified services in exchange for a fee (not pay for days worked), thereby building-in the right of people to regulate their own work. Finally, to encourage independence and emphasis the individual ownership of their company, Rank Xerox limited its purchases of goods or services to not more than 50% of the company's output at a maximum.

Having structured the jobs, we turned to selecting the people. From considerable work with career analysts, we structured a series of tests essential to ensure that individuals had the resilience of personality to withstand the world of small business where - by contrast with the (usually) supportive work groups of the large company, - the lone entrepreneur finds rapid, sharp, negative feedback the order of the day. We are fortunate now to be able to use existing networkers in counselling sessions to intending networkers and their spouses or partners (for both should be involved in, and assert to, the change which will take place in domestic life before it happens); and we seek to identify applicants as one of three categories - those who can, and wish to, build their network skills into a full-time business with other clients : those who can network, but wish to build another skill or interest into a business (examples: a trainer who runs a sports shop, and a translator who runs a farm) : and those unlikely to succeed in networking, who should stay in the world of employment.

Once accepted by line management for networking, the volunteer begins a detailed (and not cheap) training and counselling programme, receiving taxation and financial advice; sharpening up of the specific skills he or she will sell outside; development of business skills (for example, marketing presenting a case to financial backers, pricing practice); and training in the use of the telecommunicating micro. The physical environment for work at home is also important, and two alternative design schemes were prepared for the use of networkers by outside consultants - one modifying a spare bedroom into a home office, and one for those with restricted space using a series of wall-mounted units to house the microcomputer and its telephone link.

Support is necessary once the individual leaves the parent company for loneliness is a frequently - underestimated part of the small business proprietor life. To minimise this isolation, Rank Xerox takes action in two ways. First, networkers continue to be identified as part of Rank Xerox - they are on circulated lists, listed in the telephone directory, invited to department meetings and social functions, and - of course - come into Rank Xerox offices on average for a half-day a week (or one day a fortnight) to give progress reports, be briefed for new projects, discuss equipment modifications, and (naturally) to brief themselves on the latest stage of their organisations own development. Secondly, networkers are eligible for membership of Xanadu (originally) acronym for the Xerox Association of Networkers and Distributed utilities the association of individuals who have left Rank Xerox to start their own business. This association of suppliers (actual and potential) to Rank Xerox is independent of Rank Xerox, and has three fundamental objectives - to exchange business 'leads', information and services; to act on a group purchasing body; and to form a communication link between Rank Xerox and the Xanadu Companies for the dissemination of information on work available for tender, etc. From small beginnings in 1982, Xanadu has grown rapidly to a membership of well over 200 companies, with a joint turnover in excess of £12 million per year; it has recently incorporated as Xanadu Ltd (Rank Xerox has no shareholding , the company being entirely owned by its members) with a subsidiary Xanadu Consultants Ltd.

We have discussed the networkers perspective, and it remains to examine the issues involved in managing networking. Perhaps the first concern of many managers is how a limitation of a fifty percent purchase on the networking company's services can equal the networkers previous output as an employed person. The answer to this is fourfold. First, individuals are more productive working for themselves, in our experience by an increment of around 15%. Secondly, buying in a service demands tighter job specification and improved foward planning by management, and this gives a further 15% enhancement. Third, networking can be used as an acceptable, quantifiable challenge for work demands. (It will cost £3000 to do this analysis...) in a way in which costing permanent staff time is never as acceptable or as crisp in the response; this we find to give around a 30% enhancement. Finally, the enhancement of productivity due to allocating incremental roles to support staff (resulting from the reallocation of some of the networkers' former continuity mode duties) results in a further major gain of some 30% to the organisation. Obviously the factors weight differently in individual cases; but overall the output gained is not dissimilar to the previous position.

We found also that loyalty, while not precisely measurable, changed over time in nature (it became the loyalty of a small firm to a major account client, rather than the loyalty of employee to employer) but did not diminish in quantity; and that networkers were no less flexible in meeting changing management demands than permanent staff, motivated perhaps by their drive to secure further business.

In terms of contractual issues, the basic legal document is a formal supplier contract between Rank Xerox and the networking company, with the services to be provided specified in output terms with a fee for each and quality standards and timescales defined. There are minor variations for networkers - definition of health and safety responsibilities for use of the home-based micro, and agreements governing data, security are obvious ones. The price for the contract is negotiated in much the same way as with any supplier, both parties having knowledge of competitive rates for the job and and assessment of the value of the contract to each other : the eventual fee is settled by normal negotiating techniques. In terms of duration of the contract, our experience is that anything less than a one-year contract is unlikely to bring the parent company an adequate return on its training and equipment investment, while anything beyond three years is likely to be too far ahead to draw up a sufficiently detailed specification of duties.

The technology of the communication link is quite straightforward, although we would emphasise that the networkers, as professionals or executives, are very unlikely to work at the keyboard in a real time or telecommunicating mode for much of the time. It is much more probable that they will download data, model it locally on their micro, prepare their report and transmit it back, so that their telecommunications costs are negligible . The original system employed microcomputers communicating with each other over public telephone lines using an acoustic coupler and a simple software package; the more developed system employs homebased microcomputers communicating to an office based local area network (Ethernet system) to which are connected word processors, professional workstations and high quality laser printer; and current experimentation focusses on testing a wide variety of equipment from electronic typewriters to professional workstations, for working from home bases. The point in all these cases is that the technology itself is not new - the cheapness of technology is.

Any experiment has its failures, and one area in which we went wrong was in paying insufficient attention to the core staff. In placing emphasis on preparing the networkers, we underestimated the changes in the jobs of the core staff "back in the office", whereas we should have appreciated that we were in a total organisation development exercise in which both groups were equally important. Core managers, for example, had to have new skills or sharpening of existing ones for the field of designing and scheduling work, setting quality standards, and broking (or purchasing). Support staff - the secretarial/clerical/ administrative level - had new horizons opened to them, for the move of their former managers outside the organisation along with the arrival of more powerful L.A.N-based workstations, gave them both impetus and opportunity to develop as executives. A major training and career counselling need is apparent in this area, and with the involvement of the support staff themselves we have framed an appropriate development programme to meet it. As a minor aside, the "baby gap" may no longer be a career issue for female staff - an option offered on a trial basis at Rank Xerox Headquarters is to take the equipment home and remain on the payroll (but working remotely during this period).

Prospective problems which did not become issues were four. Lack of entrepreneurial capability due to big company employees having selected themselves out of the small business area did not seem a major issue, although this is not to say that everyone can be an entrepreneur. Failure of networkers' businesses is something of a joke in Rank Xerox, since success is rather more of a problem - especially for core managers negotiating with networkers! No networkers business has currently failed, and some have succeeded spectacularly; however, we must anticipate some failures, as it is not given to human beings to pick permanent staff 100% successfully and it is therefore not to be expected that a perfect selection will be made of networkers. Isolation has not yet seemed to be a problem, although the media interest in the project may have affected response on this; and technology also does not seem to have presented major issues, beyond problems due to noisy telephone lines and changes in machine procedures.

By contrast, the problems which remain to be solved are the unpredicted, and relatively unexciting, ones of defining output and quality standards (an endless debate, especially in a rapidly changing environment); and of paying attention to core staff development, structure and motivation. It is on these issues, and on testing new technology for work at home, that research currently concentrates.

In the company of the future, we predict a core group of managers and support staff which will still be quite large, but with a growing number of satellite workers either individual (networkers) or in groups (entrepreneurs). The "office of the future" may well not be an office - it may resemble a cross between an office and a community centre in a small town, and we are researching such 'neighbourhood office" concepts and is certainly not futuristic - the technology has been around for years; the important point is that it is now cheap, and any business can join in now! Within this telecommuting society, "going to work will be much less relevant a concept than "working", and many new types of organisation will arise (of which networking is but one) which will meet peoples drive for self-regulation and probably pay fees for output rather than wages for input. People, in this society will have many different roles (some waged, some fee-based, some unpaid) rather than one single, simple job - each will have what Professor Charles Handy has termed a "work portfolio" and be seeking success defined more in individual than in organisational terms. The development to this society will be gradual; the limits are not technological but lie in the identification, development and managing of people, whether core staff, networkers or entrepreneurs, in this world of the future - and this is little more than the basis task of any manager in society today.

THE ORGANIZATIONAL DEVELOPMENT

OF TELEPROGRAMMING

by Dr. Wolfgang Heilmann

INTEGRATA GmbH, D-7400 Tübingen

Contents:

1. INTRODUCTION

2. CRITERIA FOR THE SELECTION OF TELEPROGRAMMERS
    2.1 Professional selection criteria
    2.2 Personal selection criteria

3. COUNSELLING, INSTRUCTION, TRAINING
    3.1 The necessity of psychological counselling
    3.2 Poor opportunities for beginners
    3.3 The specific significance of teleprogrammers' further professional education
    3.4 Behaviour training for employees, executives and family members

4. COMMUNICATION, INFORMATION and CARE
    4.1 Intact formal communication
    4.2 Reduced informal communication
    4.3 The day at the office as a rule
    4.4 Personal guidance as an exception

5. MOTIVATION OF TELEPROGRAMMERS
    5.1 Motivation through task
    5.2 Motivation through professional development
    5.3 Motivation through situation

6. REMOTE SUPERVISION
    6.1 Reasons for a specific management style
    6.2 Selection of suitable executives
    6.3 Development of management style

7. FOOTNOTES

1. INTRODUCTION

The following exposition is based on the empirical analysis carried out by the author during the years 1982 - 1985. The sample consists of firms in the United States, England (UK), the Federal Republic of Germany and Switzerland. 32 companies with a total of 1347 teleworkstations for programmers were questioned. They mainly comprised software-houses and computer manufacturers as well as large enterprises in industry, transportation and the banking sector.

The examining procedure was characterized by standardized interviews. Discussions with experts and written inquiries played a subordinate role. Apart from the questioning of the 32 company representatives, the interviews with 37 teleprogrammers proved to be a fertile source of information. The interview results with regard to promising measures for the successful installation of telework will be presented and interpreted in the following (1).

For the successful outcome of telework several measures in personnel development (selection, judgement, formation and promotion of employees) are necessary. The structural and procedural changes of interaction-patterns that are connected with the introduction and testing of a new form of work-organization imply a "change of problem-solving behaviour of people in organizations" (2). With this personnel development becomes organizational development which pursues two targets: "the heightening of the organizational effectiveness and of the degree of correspondence between organizational practices and personal goals of the organization's members" (3).

Organizational development is not identical to organizational design. Whereas the latter comprises measures controlled by the management (4), organizational development uses, in principle, educational methods (5) and is marked by voluntariness and participation, without, however, becoming non-committal or vague. For telework, an OD-strategy, consisting of a mixture of marketing and participation (6), is advisable.

2. CRITERIA FOR THE SELECTION OF TELEPROGRAMMERS

The hiring policy of an enterprise is the first of the measures indicated. It should guarantee the personnel basis for accomplishing the company's objectives. When applied to teleprogramming, this means selection of employees already involved in the organization, since the introduction of telework usually implies the reorganization of already existing departments.

Which are the qualities potential teleprogrammers are expected to possess? Are they mainly professional qualifications or do individual characteristics play an equally important or even dominating part?

2.1 Professional selection criteria

1: Compared to other activities that are executed in telework, programming holds a special position because it bestows the employee with a high " b a r g a i n i n g - p o w e r " (7). The situation on the labour market on the one hand and professional ability on the other are shifting the bargaining power in favour of the employee desirous of home-work, so that the starting position for programmers deviates substantially from that of other professional groups.

2: The requests for high professional qualification, in particular those for experience, are numerous. As far as teleprogramming is concerned, Dorothy Kunkin-Heller is quite representative when requiring at least three years of p r o f e s s i o n a l e x -
p e r i e n c e (8); Steve Shirley not only asks for four years' professional experience but she also gets it (9). Likewise, within the empirical analysis it could be noted that the requirements set regarding professional experience of teleprogrammers are high (see table 1):

country \ Experience	USA	UK	FRG/CH	Sums
as office programmers				
- months	316	336	1634	2286
- persons	8	5	21	34
- months/person	39,5	67,2	77,8	67,2
as teleprogrammers				
- months	280	447	477	1182
- persons	8	5	21	34
- months/person	35,0	89,4	21,5	34,8

Table 1: Professional experience of teleprogrammers interviewed in months (1)

3: The fundamental reason for the great importance of professional experience as a selection criterion for teleprogrammers may, however, reside in the greater i n d e p e n d e n c e of experienced employees. This is due to the self-discipline (10) and self-organization necessary for doing telework. This becomes evident in a teleworker's account of her work at home:

- "to recognize which documents are needed,
- to initiate and supervise the necessary steps at an early stage,
- to carry out duties according to urgency,
- to divide work into those tasks that can be done quickly (in the afternoons) and those tasks that require longer working hours (in the evenings and on weekends),
- to structure work in such a way that conclusions are reached before long interruptions are made" (11).

4: The employee's organization of his time and work is especially important in teleprogramming. His ability to organize himself goes above and beyond the professional criteria, since this ability does not always develop in the course of the long years of professional experience, but also depends very much on the personality of the employee. Judkins/West, in a similar context, say: "... we have found the most intriguing issues arising from our practical experience of networking to be, not contractual, organizational or technological, but rooted in the psychology of the individual prospective entrepreneur... " (12).

2.2 Personal selection criteria

In the study of the criteria, according to which teleprogrammers are selected, the professional qualifications are almost of secondary significance. Thus programming experience was mentioned eight times and knowledge of organizational business management only four times as an essential selective criterion. More important was the character of the employee (mentioned eleven times), and the immediate needs of the firm, including the desire to keep valuable employees (mentioned nine times); see table 2:

Country (number of companies) / Criteria	USA (8)	UK (3)	FRG (17)	CH (4)	Sums (32) abs.	rel.,%
1. Programming experience	-	2	3	3	8	15,4 %
2. Knowledge of economical and organizational business	1	-	1	2	4	7,7 %
3. Character suitability	3	2	3	3	11	21,2 %
4. Family circumstances (incl. handicaps)	2	1	3	1	7	13,5 %
5. Internal needs of company (maintenance, personnel shortages)	2	-	5	2	9	17,3 %
6. No details given	4	-	9	-	13	25,0 %
Total	12	5	24	11	52	100,1 %

Table 2: Criteria for the selection of teleprogrammers (13)

1: The relatively frequent mention of character suitability is underscored by corresponding statements in the appropriate literature. Most frequently mentioned is the employee's t r u s t w o r t h i - n e s s, that is the mutual trust and confidence that must prevail between the employee and the executive when delocalization is considered (14). This trust has firstly to do with the loyalty of the teleworker and his identification with the company's objectives (15) and secondly with his industriousness, rendering constant supervision superfluous. One way to measure the trustworthiness of an employee is, in general, the number of years with the firm. The latter was used for selecting the teleworkers of Control Data (CDC) (16). Many firms, including IBM, only take "exempt" workers into their "homework program", i.e. employees with the special status who, apart from being specialists, enjoy the particular confidence of their superiors and are not paid for overtime (17).

2: During the course of the empirical analysis   t a s k - o r i e n t a t i o n   proved to be as important as trustworthiness. Teleprogrammers can be classified as intrinsically motivated, task-orientated people. This characteristic which has been found out in research (18) is a further suitable criterion for the selection process. A person who is fascinated by the job assignment itself and gives it his full attention performs well without being constantly supervised (19). The effectiveness of the work accomplished will further increase in a surrounding having fewer distractions than is normal in central offices and it will motivate the employee to work with continuity. This self-motivation can hardly be replaced by organizational measures; hence "self-starters" are "good candidates for remote work" (20). They motivate and discipline themselves and do not need continuous communication with executives and colleagues. On the contrary, this could be demotivating, for they are people "who like to be alone" (21).

3: The last remark describes the little   n e e d   for   s o c i a l   c o n t a c t s   of potential teleworkers. For such workers, isolation on the job is not experienced as a burden, but as a means of achieving job satisfaction (22). The assumption that problems are more easily solved by contemplating a series of interdependent steps and - like the job of programming - are done better by individuals in isolation (23), well complements the assumption that programmers tend to be loners (24) and explains the behaviour of many teleworkers. Task-orientation and little desire for social contacts therefore turn out to be important selection criteria for potential teleworkers.

3. COUNSELLING, INSTRUCTION, TRAINING

3.1 The necessity of psychological counselling

1: If the criteria for the selection of teleprogrammers are neglected, considerable disturbances in organization as well as in the individual could occur. Discrepancies regarding independence and other people-oriented criteria are particularly problematic. Professional skill and experience can be fairly easily judged; the assessment of attributes such as job-organization and the need for social contacts, however, require   p s y c h o l o g i c a l   t e s t s .   These are actually frequent sources of misjudgements, but they nevertheless do give valuable information.

2: A more serious problem than misjudgement by others is an incorrect   s e l f - a s s e s s m e n t . Due to this millions of people unhappily perform jobs daily for which they are not suited. Obviously they do not know the real reason for their frustration, because they are too involved in their situation in order to reflect objectively about it. In view of this, career counselling appears to be a necessary institution in a society striving for self-fullfillment.

3: With regard to the organizational structure of teleprogramming the situation must be judged similarily: before the installation of a remote work-place is decided upon,   p s y c h o l o g i c a l   c o u n s e l l i n g   should take place. Moreover, it appears necessary to promptly discuss the particular technical and organizational aspects of telework with the workers concerned. A prerequisite for a productive discussion is an elaborate job description that - apart from the classification of the job and the arrangement of the communicative relationships - describes the most important tasks and expectations (25). Only then can the specific expectations be derived and a job-profile drawn up. On the basis of such profiles can it be determined whether or not an employee is suited for decentralized work, and only after positive results of the counselling can instruction and training of the persons suited for telework be considered (26).

3.2 Poor opportunities for beginners

1: Generally the teleworker's training opportunities are more limited than those of office workers because the teleworker is not as available for "on the fly" training as his colleagues in the central office. Moreover, research done by the state of Baden-Wurttemberg shows that new teleworkers reach a productivity level inferior to that of experienced employees and also learn more slowly at the decentralized work-place than in the central office (27). It appears that, apart from formal instruction, the incidental o n - t h e - j o b - t r a i n i n g is also very important for developing knowledges (28). For this reason, the introduction to the tasks is made more difficult for the beginner.

2: The supposition that beginners in the profession can be employed as teleprogrammers is, however, not completely wrong, as has been proven by the examples given by Lift in Chicago and PSG in Munich. By means of a thorough theoretical and practice-oriented instruction both institutions successfully prepare handicapped beginners to perform as programmers. In the case of Lift, the candidates are subject to a severe selection procedure, survived by only one of twenty applicants. This is succeeded by a six-months, mainly audiovisual training period and, finally, a one-year trial period before a definite assignment with a contracting firm can be achieved. At the time of inquiry (summer 1982) 19 out of 44 trained persons had survived the procedure; many had already been active as programmers for years. The instruction at PSG takes somewhat longer (18 - 36 months).

In general, i.e. outside of specialized institutions, beginners have only a slim chance of taking over qualified jobs. The normal way to qualification as an independent programmer takes a long time and this time-period is necessary in order to achieve recognition as teleprogrammer.

3.3 The specific significance of teleprogrammers' further professional education

1: Further professional education for teleprogrammers must be distinguished from that of office programmers with regard to nature and extent. During our research, the n e e d   t o   f u r t h e r   e d u c a t e   teleprogrammers could be determined (see table 3).

An analysis of this table shows that teleprogrammers participate in professional education programs five days per year on the average, according to the persons concerned, whereas according to the executives' statements, the time spent for further education is somewhat greater, but still lies far below the average of 10,9 days that was determined for the Federal Republic of Germany (29).

Further professional education in days/year	USA		UK		FRG/CH		Sum of answers	
Country	EX (30)	EM (31)	EX	EM	EX	EM	EX	EM
0	2	8	1	-	3	6	6	14
1 - 5	-	2	1	3	1	2	2	7
6 - 10	1	-	1	1	8	6	10	7
more than 10	1	-	-	-	4	6	5	6
no details given	4	1	-	1	5	1	9	3
Total	8	11	3	5	21	21	32	37

Table 3: Extent of further professional education of teleprogrammers, in days/year

2: The extent of i n d e p e n d e n t   s t u d y of literature, brochures and similar documents supported by the companies is apparently of no great importance, since only 13 out of 37 teleprogrammers made such statements. Even though it could not be ascertained how much time a teleprogrammer invests in independent study, it can nevertheless be assumed that, in general, a negative development is taking place, which - in the long run - could lead to the loss of the qualification-advantage over office programmers and finally to dequalification.

3: In order to stop such a development, considerably greater efforts need to be made, even greater efforts than those made with office programmers. The latter are, in fact, confronted with the same rapid technological advances as are teleprogrammers; however, their constant contact with colleagues facilitates keeping informed (32). To compensate for this deficit of spatial separation from the central office, s p e c i a l   p r o g r a m s   f o r   f u r t h e r e d u c a t i o n should be developed for teleprogrammers. Not training measures by means of video cassettes or computer supported instruction in the sense of teletraining (33) should be initiated, but central, personal meetings with colleagues and superiors (34).

3.4 Behaviour training for employees, executives and family members

In addition to professional education, totally new forms of training are equally necessary to promote personal committment. Firstly, behaviour training is a much greater task than merely imparting knowledge and secondly, failure to offer such training would have negative consequences even on very capable employees. This is why a telework program must include a training program that facilitates the employees' start in a new form of organization and makes it easier for them to remain with it.

1: Behaviour training for teleworkers should be looked at as part of the psychological and technical-organizational counselling and is best organized by way of on-the-job-training during the initial training phase. The employee must learn behaviourial patterns which facilitate contact with the central office on the one hand and working at the decentralized workplace on the other.

2: The teleprogrammers' c o l l e a g u e s   and   s u p e r i - o r s should be included in this training, in order to find out which behaviour patterns are best for achieving the organization objectives. In the following procedures of information, communication and guidance must be set up and tested for their usefullness.

3: Inclusion of f a m i l y   m e m e r s   in behaviour training is necessary only then when the workplace is at home, not however, when satellite- or service offices are involved. Since a work-place at home is a lasting intrusion into the family life one may not simply take its implementation for granted. Instead, it is necessary to provide the husband or wife and the children with the opportunity to become acquainted with the new situation and also participate in it (35).

4. COMMUNICATION, INFORMATION and CARE

Programming is one of the activities that actually involve a great deal of information, but requires relatively little personal communication. But, within the organizational development, the wide field of informal communication processes must be observed apart from formal communication, the former being essentially defined by the sort of persons involved. Due to the assumption that programmers and teleprogrammers in particular are intrinsically motivated, task-orientated people, it can be expected that decentralization will lead to relatively little withdrawal symptoms.

Nevertheless, special regulation of information and communication for teleprogrammers appears justified because, unlike with office programmers, they cannot count on normal verbal information exchange during their daily work, but are confronted with problems caused by distance.

4.1 Intact formal communication

The formal process of information exchange rarely gives reason for complaint by teleprogrammers. Surveys indicate that factual information deficits barely exist, or at least are rarely experienced as such:

Country	USA		UK		FRG/CH		Sums of answers abs.		Sums of answers rel. (in %)	
Information deficiencies	EX (30)	EM (31)	EX	EM	EX	M	EX	M	EX	M
rare 1	3	3		1	2	2	5	6	15,7 %	16,2 %
2		2			6	6	6	9	18,8 %	24,3 %
3		1	1	2	3	6	4	9	12,5 %	24,3 %
4		2				1		3		8,1 %
5		2		1	2		2	3	6,3 %	5,4 %
6				1	1	1	1	2	3,1 %	5,4 %
frequent 7										
no details given	5	1	2		7	5	14	6	43,4 %	16,2 %
Total	8	11	3	5	21	21	32	37	99,8 %	99,9 %

Table 4: Frequence of information deficit

The first three assessments (1 - 3) on the seven-piece interval scale, that indicate a rare appearance of information deficits, made up 64,8 % of the answers given by 37 of the interviewed teleprogrammers and 49,0 % of the 32 executives, 43,4 % of whom, however, did not express their opinion. The amount of answers that expressed frequent appearance of information deficits (i.e. the last three assessments 5 - 7) only totalled 10,8 % of the teleprogrammers and 9,4 % of their executives. It may therefore be assumed that the factual information exchange practised within teleprogramming functions as well as in office programming which is also normally accompanied by certain amount of information deficit during daily work.

4.2 Reduced informal communication

The informal process of information exchange and formation of opinion within the company, however, must be judged differently, as was explained by a woman-teleprogrammer: "With respect to ... factual communication no particular problems arose after an initiation period. I could always obtain missing information by telephone or in writing. In cooperation with one of the secretaries, communication via terminal is possible and successful. Communication of office events is reduced, however, as could be expected. The same is true for participation in the formation of opinions in company matters, for information on other projects and for the knowledge of current matters of interest in the firm" (36). Hence, a reduction in the number of personal conversations does not, in general, lead to disturbances in work, but certainly to a deterioriation of the informal relationships. The feeling of being neglected by the central office must therefore be interpreted as an expression of the teleworker's need for contacts and not so much as a general weakness of this organizational form of work.

4.3 The day at the office as a rule

1: The last observation is valid for a quite definite situation that is said to be typical based on the information that teleworkers go into the office occasionally in order to inform themselves about office happenings and to communicate with executives and colleagues. The following table gives information on the f r e q u e n c y and d u r a t i o n of this presence at the office, gathered in a survey of 37 teleprogrammers:

Frequency of office visits / Duration of office visits	less than 1 hr.	one hour	more than 1 hr.	no details given	Sums abs.	rel., %
less than 1 times per year				1	1	2,7 %
less than 1 times per month			3		3	8,1 %
once a month			1		1	2,7 %
several times per month	1		6		7	18,9 %
once a week			7		7	18,9 %
several times per week			9		9	24,4 %
once daily			6		6	16,2 %
no details given			1	2	3	8,1 %
Total absolute	1		33	3	37	
Total relative	2,7 %		89,2 %	8,1 %		100,0 %

Table 5: Frequency and duration of presence at the office

The table makes clear that the majority of the interviewed persons was regularly present at the office as often as several times per month. Aside from the employees who were present daily and only did telework occasionally after closing time or on the weekend, they amounted to 62,1 %. Only 10,9 % came to the office less than once a month. The average duration of office visits in 89,2 % of the cases was several hours. From this another rule for the formation of teleprogramming can be derived: the set-up of a regular day spent at the office.

2: Due to the empirical analysis as well as organizational demands it appears necessary to reserve one office day per week for teleprogrammers (37). Justification for exactly o n e   w o r k i n g   d a y can, however, not be given, since the extent of the information exchange required at any given time varies respectively according to the job-assignment, the number of items to be communicated and according to team-size. As a rule, 10 - 20 % of worktime should, however, be enough and normally is sufficient. Merely in those cases in which the visit to the office must also be used for doing certain activities (like testing and implementing) the time spent must be prolonged accordingly, so that several office days per week may become occasionally necessary (38). Communication problems then prove to be considerably less significant than they would be if teleworkers came to the office less regularly.

4.4 Personal guidance as an exception

1: In the literature on the subject of telework the demand for personal guidance or supervision of the teleworker is occasionally expressed. The aim is, of course, to reduce isolation or to do away with it entirely (39). For example, the management of F. International spends a lot of time communicating with the employees by telephone (40), in particular with those who momentarily cannot be employed (41). For this reason a "staffing manager" is appointed who is not identical with the project manager or department head. Generally, this type of care is not of great importance.

2: Our s u r v e y , instead, revealed the following (see table 6). This table shows that 54,0 % of the teleprogrammers were looked after by their project managers or department heads and that in only 16,2 % of the cases another person personally in charge of them was mentioned. 29,7 % gave no comments on this subject or they suggested that they considered such care unnecessary.

Care by \ Country	USA EX (30)	USA EM (31)	UK EX	UK EM	FRG/CH EX	FRG/CH EM	Sums of answers abs. EX	Sums of answers abs. EM	rel., in % EX	rel., in % EM	
project manager		3	1	3	6	8	7	14	21,8 %	37,8 %	⎫
other superior	2	3			3	3	5	6	15,6 %	16,2 %	⎬ 54 %
other person		2	1	2	1	2	2	6	6,2 %	16,2 %	⎭
no details given, or rather not necessary	6	3	1		11	8	18	11	56,2 %	29,7 %	
Total	8	11	3	5	21	21	32	37	99,8 %	99,9 %	

Table 6: Caring for teleprogrammers

3: Taking into consideration that in two of the most important cases in which personal care of teleworkers was required, or even put into practice - i.e. with F. International and Rank Xerox in London - freelance workers on a   p a r t t i m e - b a s i s   were involved; this, together with the fact that their jobs were not guaranteed, shows the exceptional character of this regulation. Personal care of full-time teleworkers beyond the weekly visit to the office does not appear to be necessary.

5. MOTIVATION OF TELEPROGRAMMERS

"Motivation ensues from the interaction between a person and a situation" (42). The measures aimed at motivating a person should activate the incentives that are inherent to a particular situation and make them clear to the employees concerned. Teleprogrammers are intrinsically motivated workers or, rather, should be chosen from a group of co-workers who are of such a predispositon. Hence, their motivation grows mainly from the incentives inherent in this kind of work, which, therefore, must remain in the centre of the endeavours. Only secondarily should those incentives be included that arise out of the external situation of an employee.

5.1 Motivation through task

The motivation potential inherent in work is operationalized according to Hackman/Oldham by the factors skill variety, task identity, task significance, autonomy and feedback (43) and by using these factors different organizational forms of work can be compared. The following is a survey which compares telework to office work for programmers:

- With regard to versatility, teleprogramming is more motivating than office programming, because it is, in general, of greater variety.

- Teleprogramming is more strongly oriented towards an identifiable result than is office programming. For this reason the identity factor plays a larger role.

- The significance of teleprogramming for the teleworker or for other people does not differ from the significance of office programming, because the results remain the same.

- Autonomy, i.e. the extent of independence while working and the responsibility for success and failure is greater with teleprogramming.

- Only the feedback from colleagues and superiors may be better guaranteed in office programming and hence be better motivating than in teleprogramming.

All in all, it may therefore be concluded that teleprogramming has a higher motivation potential than office programming. Consequently, special efforts to motivate intrinsically motivated teleworkers are not necessary.

5.2 Motivation through professional development

1: The professional development of the individual worker, i.e. his career, is also closely related to the content of the work. As Schanz quite rightly affirms, this does not only mean promotion into executive positions, but also the assuming of other responsibilities in staff positions or at important coordinating points in the hierarchy (44). Many people do not strive for a career in the usual sense, but for the further development of their own capabilities and for the pure challenge of tackling technically sophisticated job-assignments. It is therefore necessary to interpret the concept "career" neutrally and to differentiate between the various "career anchors" (45). The " c a r e e r   p a t t e r n " of programmers as intrinsically motivated, task-orientated people may be characterized by the following order of "career anchors":

- The technical-functional competence of an occupation takes the first place in an individual's assessment. "This in turn guides the ambitions of those individuals who appreciate the challenges of their special field. A rise within the organizational hierarchy is attractive only when it takes place within one's own functional sphere" (46).

- The creativity component, i.e. the need to develop something new, has for many programmers a very high priority. This behaviour component leads in many cases to technically brilliant inventions which sometimes results in the entrepreneurial founding of small enterprises for marketing one's own know-how.

- In another group of programmers the need for security dominates and suppresses the desire for independence. Such employees value a secure and stable job, a good income and attractive social welfare benefits.

- The management component must be considered a motivating factor as well: with programmers, however, it mainly exists as the desire to take over a technically qualified executive job, for example as project manager.

- Autonomy and independence are certainly much more sufficient in doing remote work than in a centralized situation.

2: The intention to motivate, therefore, presents itself as follows: above all one must guarantee that teleprogrammers are given a challenging job through which they can grow and develop professionally. For a certain group of these employees   p r o j e c t   m a n a g e r   p o s i t i o n s   must be readily available. Since in reality project managers have been sought after very much for years, and since this function can be executed in telework, motivation derived from the possibility of advancement is very possible and is frequently practised. Our investigations prove that among the status changes that take place during employment as teleprogrammer, the promotion to the project manager is the most frequent one (see table 7):

Country / Status	USA t_1 (47)	USA t_2 (48)	UK t_1	UK t_2	FRG/CH t_1	FRG/CH t_2	Sums t_1	Sums t_2	Δt (49)
1. Assistant programmer or unemployed	1				1		2		− 2
2. Programmer	1	2	1		4	3	6	5	− 1
3. Chief programmer			2				2		− 2
4. Systems programmer	1				1	1	2	1	− 1
5. Systems analyst	2	2	1	1	4	1	7	4	− 3
6. Project manager		1	1	3		2	1	6	+ 5
7. Consultant	2	2				2	2	4	+ 2
8. No details given	4	4		1	11	12	15	17	+ 2
Total	11	11	5	5	21	21	37	37	0

Table 7: Status alterations of teleprogrammers interviewed

5.3 Motivation through situation

1: People with an aptitutde for motivation favourable to telework do not need to be motivated to take over work at a decentralized work-place. A high percentage of employees are, however, of a different nature including many programmers. Which circumstances motivate them to work at home and how can this motivation be s t a b i ‑ l i z e d ? The following example makes this point clearer:

One of the teleprogrammers interviewed was interested in team-work and communication with professional colleagues. She classified herself as a person who did not want to remain in a quiet corner, but was intent on making a career of her work, by means of promotion, more influence and prestige. Her absence from the office for family reasons appeared to her to be a threat to her career chances and was accepted by her only for a short period of time.

2: Teleworkers of such a motivation structure may be t e m p o ‑ r a r i l y interested in working at home because of situational factors such as family obligations, physical handicaps or long distances. In most cases, they will react positively when telework is the only alternative to unemployment; they will, however, remain basically dissatisfied because at home they lack extrinsic motivation. Aside from the situations in which habit is stronger than the need for communication, or when inflexible conditions prevail, remote work should be kept to a minimum. Even people who because of handicaps can be transported only with great difficulty must be regularly brought to the office, if they are of an extrinsic motivation structure, so that they can renew their contacts there (50).

3: I n   c o n c l u s i o n   it may be established that with respec to the motivation of teleprogrammers no problem will arise after selec ting the right people. However, motivationally unsuited employees wil hardly be persuaded to work at a decentralized place much longer than necessary. Motivation due to personal situation is therefore effective for a certain amount of time only. From this it can be concluded that pressure to do work at home should not be exerted, because this would lead to frustration in certain people. A positive alternative to the decentralized work location would be satellite or services offices which, as "easily, firmly structured small-scale groups" develop the "highest performance motivation" (51).

6. REMOTE SUPERVISION

Following the reflections on whether employees are qualified for the organizational form telework or not, the discussion now turns to the measures that will enable executives to adjust to the new situation, with emphasis lying on developing an adequate management style.

By "management style" is understood a "uniform and constant behaviour pattern" which is marked by a certain basic philosophy (52). This fundamental principle is based on the executive being either task- or people-oriented. It can be shown that task-oriented managers face telework more open-mindedly than the people-oriented ones (53). This has been determined from the analysis of individual cases of teleprogramming and is confirmed indirectly by relevant literature. There it is stated that a special management style is required which perhaps does not suit all managers, but deals with the fact that the employees work at remote locations. "Remote supervision" is said to be necessary (54). The most important criteria for this "tele-management" are summarized as follows:

- central work distribution
- self-organization of teleworkers
- methodical mode of work
- regular control of results.

Hence it is a question of a more or less clear development form of management-by-objective (MbO), so that it suffices here to have mentioned the inherent objectification of management style as compared to traditional forms of management (55).

6.1 Reasons for a specific management style

Arguments for a task-oriented management style can be seen in the fact that three out of four management qualities mentioned by Gibb (56) are specifically influenced by telework:

- The work situation of teleprogrammers which is characterized by the kind of job and the state of technology hardly permits spontaneous face-to-face contact. Instead this must be replaced by scheduled meetings and by communication via technical mediums.
- The personal relationship within the work groups and their system of norms experience a not completely foreseeable but definite loosening of personal contacts among colleagues due to spatial separation.
- The attitudes, desires and problems of the employees who are, if correctly selected, mainly intrinsically motivated, suggest a task-oriented management style.

The fourth component of efficient management, the personality of the executive, should do justice to the particularities of managerial work mentioned above. Measures aimed at influencing the personality of the executive in order to objectify his management style can be characterized according to whether they aim at selecting or training the executive.

6.2 Selection of suitable executives

1: At present, the selection of suitable executives is of subordinant practical relevance, since the event of teleprogramming is as yet limited. It is of certain importance in as far as it is recommendable for the implementation of decentralized projects to involve only such executives who are interested in this work and thus are suitable for it. "If the requirements of remote work make a manager uncomfortable for any reason, the manager should not permit it because it simply will not be successful", the research report of Diebold states (57).

2: In the long run, assuming telework becomes more common, the task-oriented personality must be considered when hiring executives. C r i t e r i a for measuring leadership behaviour, and consequently for selecting suitable executives have been developed within management diagnostics. Since we are not searching for generally valid selection criteria, but for those valid specifically for telework, the chances of success of the selective process may be judged favourably. The left side of the following list of aims should be emphasized without neglecting those to the right (58).

- achievement of goals
- locomotion
- professional cooperation
- instrumental leadership
  .
  .
  .

- group preservation
- cohesion
- personal cooperation
- social-emotional leadership.
  .
  .
  .

Task-oriented management personalities, who are able to lead their employees cooperatively without needing continuous personal communication, are called for.

6.3 Development of management style

1: The growth of already active executives is, at least in the short run, more important than the selection of applicants. It also is more likely to do justice to the fact that people do not possess static qualities (59), but that they show characteristics that develop dynamically. Modification of behaviour through instruction, further education and training is not only possible, but is the usual way people adjust themselves to a new undertaking (60).

2: In remote supervision, techniques like conferences, frontal teaching, self-instruction with the help of books or programmed teaching are, however, relatively ineffective. When behaviour is to be altered, p r o c e s s - o r i e n t e d   t e c h n i q u e s like role-playing, case studies and development according to the mangerial grid have greater chances of success (61). Therefore, it is necessary to refer to the behaviour training for employees which, as mentioned, should be carried out together with the executives by way of on-the-job-training and team training (see section 3.4).

3: R e g a r d i n g   c o n t e n t, the behaviour training of executives must be oriented to the same goals that have been described as being important for selection and that were presented separately during the discussion of management techniques. In principle, it can be assumed that executives who are not exclusively people-oriented have a good chance of learning tele-management, "but some managers won't be able to cope without close supervision" (62).

7. FOOTNOTES

(01) Further details and sources can be taken from the monography: *Teleprogrammierung - Die Organisation der dezentralen Software-Produktion*, published by the same author within short in the Forkel-Verlag, Wiesbaden.
(02) Bartölke, K., *Organisationsentwicklung*, in: Grochla, E., Handwörterbuch der Organisation, 2nd edition, Stuttgart 1980, column 1469.
(03) Ibid 1469/70; similarly Kieser, A. et al.,*Organisationsentwicklung*, in: Organisationstheoretische Ansätze, Munich,1981, pp. 113 and 127.
(04) Bartölke, K., *Organisationsentwicklung*, l.c., column 1469.
(05) Kirsch., W., et al., Das *Management* des geplanten organisatorischen Wandels von Organisationen, Stuttgart 1979, pp. 70 and pp. 207, with a list of "Methods of Organizational Development".
(06) Ibid., p. 310.
(07) See in particular Olson, M., *Telecommunications* and the Changing Definition of the Workplace, Center for Research on Information Systems, New York University, April 1982, pp. 6 as well as: Impact of Information Technology on Work Organization *Position Paper 1*, International Conference of the Government of the Federal Republic of Germany in Cooperation with the OECD; 1984 and after: The Social Challenge of Information Technologies, Nov.28-30, 1984; Draft Aug.31, 1984, pp. 22.
(08) Kunkin-Heller, D., *Industry* Taps Productivity of Part-Time, At-Home Programmers, in: Infoworld, Dec.7, 1981.
(09) Ibid: My Home is My Office, in: Capital, 05/81, p. 240.
(10) See by ex. Olson, M., *Remote Office Work*: Implications for Individuals and Organizations, New York University, Working Paper Series, 1981, pp. 27 and *Changing Work Patterns* in Space and Time, in: Communications of the ACM, March 1983, vol. 26, no. 3, p. 185.
(11) Krcmar, B. in: Heilmann, W., Krcmar, B., *Formen* und Modelle der Telearbeit, lecture during the 7th International Congress of ADV, study group for data processing, "Informationstechnologie: Realität und Vision", Vienna, March 19-23, 1984, Proceedings, p. 258.
(12) Judkins, P.E., West, D., *Networking*. The Distributed Office. A New Venture in Modes of Employment. Edited by Rank Xerox Limited, 5th edition, London, Oct. 1982, p. 14.
(13) Multiple statements possible. When 13 firms made no statements this is due to the fact that they do no planned telework, but had set up the work-place concerned only as an exception.
(14) Canning, R.G., Experiences with *Tele-Commuting*, EDP-Analyzer, Nov. 1982, p. 7; Diebold Group Inc., *Office Work in the Home: Scenarios and Prospects for the 80's, New York, Aug. 1981, p. 32; Kendall, S.D., Smith Cunnien, P.J., Courseware Operations

and Plato Development. Alternate Work Site Programs: Interim *Evaluation*, edited by Control Data Corp., Minneapolis, March 1982, p. 12.
(15) Diebold Group Inc., *Office Work*, l.c., pp. 31/32.
(16) Kendall, S.D., Smith-Cunnien, P.J., *Evaluation*, l.c., p. 2.
(17) See: Heilmann, H., Heilmann, W., *Softwareentwicklung* am Telearbeitsplatz, Erfahrungen und Trends aus den USA, in: HMD, no. 110, March 1983, pp. 97/98; see also DeSanctis, G., A *Telecommuting Primer*, Some Guidelines on How to Manage a Work-at-Home Project, in: Datamation, Oct. 1983, p. 216.
(18) See Heilmann, W., *Teleprogrammierung* - Die Organisation der dezentralen Software-Produktion, Wiesbaden, 1987, part V, point 2.1.2.
(19) Olson, M., An *Investigation* of the Impacts of Remote Work Environments and Supporting Technology. Center for Research on Information Systems, New York University, Sept. 1983, p. 9.
(20) Diebold Group Inc., *Office Work*, l.c., pp. 31/32.
(21) Ibid p. 32.
(22) Canning, R.G., Experiences with *Tele-Commuting*, EDP-Analyzer, Nov. 1982, p. 6.
(23) Hill, W., et al., *Organisationslehre*, Ziele, Instrumente und Bedingungen der Organisation sozialer Systeme, 2 volumns, 3rd edition, Bern, Stuttgart, 1981.
(24) Couger, D.J., Zawacki, R.A., Was motiviert EDV-*Fachleute*? in: Büro und EDV, 12/1978, p. 29.
(25) Hentze, J., Arbeitsbewertung und *Personalbeurteilung*, Stuttgart, 1980, pp. 111.
(26) Similarly Judkins, P.E., West, D., *Networking*, l.c., p. 15.
(27) Wawrzinek, S., Dezentrale *Texterstellung* mit Teletex: First results from research done by the state of Baden-Wurttemberg "Schaffung dezentraler Arbeitsplätze unter Einsatz von Teletex"; lecture on the occasion of ONLINE '85, Febr.12-15, 1985, Düsseldorf, pp. 3N11.
(28) "Office acculturation, the subconscious absorption of working knowledge as opposed to that acquired through a direct learning process, is an informal but very important part of developing knowledge", so Raney, J.G., *Project Homebound*, p. 15, in: Office Work Stations; edited by National Research Council, Washington D.C., 1985.
(29) Kienbaum Unternehmensberatung GmbH., editor. *Vergütungsberatung*, 1982, p. 18.
(30) EX = the answers of executives
(31) EM = the answers of employees.
(32) The employee working at a remote work-place is lacking the possibility to compare her own performance with that of her colleagues; "no mechanisms are avaiable that would inform the employee at the remote work-place about her performance standard". Wawrzinek, S., Fröschle, H.P., Schaffung dezentraler Arbeitsplätze unter Einsatz von Teletex, *Zwischenbericht*; edited by Fraunhofer-Institut für Arbeitswirtschaft und Organisation (IAO), Stuttgart, 1985, p. 48.
(33) As being offered by ex. by the University of New Jersey; see New Jersey Institute of Technology, Division Continuing Education, New Offerings, fall 1983, pp. 4-8.
(34) See Shirley, St., Office *Workstations* in the Home, p. 9: "The corporation uses computers based 'open' or 'distance' learning methods but finds that conventional inhouse training has an important bonding effect, being one of the few occasions when groups meet each other away from the client-situation"; prepared for the Executive Forum of the National Research Council, Nov.9, 1983.

(35) Judkins, P.E., West, D., *Networking*, l.c., pp. 14/15 as well as Olson, M., l.c., *Remote Office Work*, p. 37.
(36) Brigitte Krcmar in: Heilmann, W., Krcmar, B., *Formen*, p.259.
(37) Similarly by ex. Holzer, Chr., Die *Nachbarschaftszentrale* - ein gutes Beispiel zur Technikgestaltbarkeit; diploma essay, Institut für Soziologie, Karlsruhe University, March 1984, p. 105., Nilles, J.M., *Teleworking* - Working Closer to Home, in: Technology Review, April 1982, p. 59. Phelps, N., Mountan Bell: *Program for Managers*, p. 35, in: Office Work Stations in the Home. Edited by National Research Council, Washington D.C., 1985. In the research report of Diebold is stated that frequency of presence is inferior to once a week. See *Office Work*, l.c., p. 28.
(38) With CDC, the typical teleworker only spends 2-3 days at home, apart from this he works in the office. See Kendall, S.D., Smith-Cunnien, P.J., *Evaluation*, l.c., p.3. Steve Shirley also reports that an employee of F. International only spends 35 % of her work-time at home, on the average. See Shirley, St., F. International *Twenty Years Experience*, p. 54, in: Office Work Stations in the Home. Edited by National Research Council, Washington D.C., 1985.
(39) Thus Rank Xerox further provides all 'networkers' with circular letters and invites them to festivities. See Judkins, P.E., West, D., *Networking*, l.c., p. 17.
(40) Shirley, St., The *Remote Control* of Projects; paper presented to the EURO-IFIP Conference 1979, p. 120.
(41) Ballerstedt, E., et al., *Studie* über Auswahl, Eignung und Auswirkungen von informationstechnisch ausgestalteten Heimarbeitsplätzen; edited by: Bundesministerium für Forschung und Technologie, Karlsruhe, 1982, p. 88.
(42) Rosenstiel von, L., Grundlagen der *Organisationspsychologie* - Basiswissen und Anwendungshinweise, Stuttgart, 1980, p. 270.
(43) Hackman, J.R., Oldham, G.R., *Motivation* Through the Design of Work, in: Organizational Behaviour and Human Performance, vol. 16, 1976, pp. 256; viz the translation by Schanz, G., *Organisationsgestaltung*, Struktur und Verhalten, Munich, 1982, p. 138.
(44) Schanz, G., *Organisationsgestaltung*, l.c., p. 252.
(45) The conception of career anchors by Edgar H. Schein, How *"Career Anchors"* Hold Executives to Their Career Path, in: Personnel 52, 1975, pp. 14; this is being included in the reflections of Schanz, G.,viz: *Organisationsgestaltung*, l.c., pp. 254.
(46) Schanz, G., *Organisationsgestaltung*, l.c., p. 254.
(47) t_1 = status before telework was initiated.
(48) t_2 = status at the time of interview.
(49) $\triangle$ = number of status alterations; + = promotion.
(50) See Murphey, A., How one Able Man Overcame *"Disabled Fears"*, in: Software News, May 1983, p.33.
(51) Eiff von, W., *Organisationsentwicklung*, personalpolitische, strukturelle sowie kostenleistungsorientierte Aspekte organisatorischer Änderungen; Berlin, 1979, p.250.
(52) Hörschgen, H., *Grundbegriffe* der Betriebswirtschaftslehre II, Stuttgart, 1979, p. 150; as to this term also see Wunderer, R. et al., *Führungslehre I*. Grundlagen der Führung, vol. I, Berlin, New York, 1980, pp. 220.
(53) See Heilmann, W., *Teleprogrammierung*, l.c., part V, point 1.2.2.4.
(54) Olson, M., *Remote Office Work*, l.c., pp. 29 and New Information Technology and *Organizational Culture*, New York University, Working Paper Series, July 1982, pp. 32; also see

Diebold Group Inc., *Office Work*, l.c., pp. 31. as well as Kawakami, S. S., Electronic *Homework*. Problems and Prospects from a Human Resources Perspective, Dissertation, Institute of Labor and Industrial Relations of Illinois at Urbana-Champaign, July 9, 1983, pp. 73.
(55) Further details in Heilmann, W., *Teleprogrammierung*, l.c., part V, point 4.4.6.
(56) According to Wunderer, R., et al., *Führungslehre I*. Grundlagen der Führung, vol. I, Berlin, New York, 1980, p. 146.
(57) Diebold Group Inc., l.c., *Office Work*, p. 33.
(58) See Rosenstiel von, L., *Organisationspsychologie*, l.c., p. 127.
(59) Tenbruck, F.H., Der Mensch als *Merkmalsträger*. Wenn die Sozialforschung die Privatsphäre veröffentlicht und zerstört. In: FAZ, Frankfurt, March 31, 1984, as well as in: Zur *Kritik* der planenden Vernunft, Munich, 1972, in particular pp. 72.
(60) Rosenstiel von, L., *Organisationspsychologie*, l.c., p. 92.
(61) Rosenstiel von, L., *Organisationspsychologie*, l.c., p.130. Also see Bartölke, K., *Organisationsentwicklung*, columns 1475.
(62) Schloßberg, J., Home Is Where the *Job* is, in: Digital Review, 10/1984, p. 49.

2

ECONOMIC AND SOCIAL FACTORS

REMOTE POSSIBILITIES: SOME DIFFICULTIES IN THE ANALYSIS AND QUANTIFICATION OF TELEWORK IN THE UK

Ursula Huws

Empirica UK
20 Canonbury Square
London N1 2AL, UK

1. INTRODUCTION

Although the word 'telework' has not yet achieved the status of a dictionary definition, it is now familiar to researchers and policymakers concerned with the future of work and even to some employers. Most of these people, it must be assumed, think that they know what they mean when they use the term, or one of its many pseudonyms such as 'telecommuting', 'the electronic cottage' or 'networking'.

In common speech there does indeed appear to be a working concensus of opinion on the subject. It seems to be generally believed that information technology has made it possible to decentralise many types of work involving the electronic processing of information, and 'telework' is simply the term used to describe workers who have been dispersed geographically in this way.

It is only when one attempts to collect empirical data on telework, so that its development can be quantified and future trends extrapolated, that this concensus dissolves. As soon as we attempt to count the numbers of teleworkers or locate the industries in which they are situated, it becomes clear that we have no stable or concrete definition of what we mean by the term. In fact it is doubtful whether it is even possible to construct one using existing conceptual and statistical categories.

In the UK, at least, no academic researcher has come anywhere near producing an authoritative estimate of the numbers of teleworkers or a coherent analysis of the tasks involved and the process by which they become transformed into telework. The working definitions of telework which are employed tend to fall into two categories. Either they are extremely imprecise or else they avoid logical definition and merely list a motley collection of different employment or organisational types which the researcher has chosen to categorise as telework. These typologies frequently lack consistency and often implicitly include categories of work which fall well outside the common speech notion of teleworking, requiring exclusion clauses in order to function effectively as definitions. Thus, for instance, we may find that a definition includes homeworkers, but not those carrying out 'traditional work', or sales representatives who have been issued with portable terminals, but only those who sometimes work from their homes. This paper argues that the most useful way to conceptualise telework is not as a unitary phenomenon but as the point of convergence between several different trends currently affecting the organisation of work.

These can be summarised under five broad headings: the
geographical relocation of employment; the externalisation of
labour; changes in the contractual relationships between
employers and workers; increases in home-based working; and
changes in the design of jobs.

The introduction of new information technology plays a
significant role in facilitating several of these trends, but in
no case is it the exclusive determinant. To understand them
fully it is also important to take account of other factors,
social, economic, legal and in some cases political.

It is also true to say that none of these trends is exclusively
concerned with what is generally described as telework, though
all could be said to embrace it. The specific phenomenon of
telework, if indeed such a phenomenon can be said to have an
independent existence, lies at the intersection of these wider
trends.

The rest of this paper draws on the UK experience to support this
proposition and illustrate some of the difficulties facing
researchers attempting to analyse or quantify telework.
Inevitably many of the conditions is describes are specific to one
country. It is not therefore necessarily possible to draw general
conclusions from it which are applicable elsewhere and extreme
caution should be exercised by anyone attempting to do so. Indeed,
until a great deal more empirical research has been done in
several countries, there are very few general statements about
telework which can be put forward with any confidence that they
can be substantiated. Let us hope that the March 1987 empirica
conference will establish a common conceptual framework which can
form the basis for such research in the future.

2.   GEOGRAPHICAL RELOCATION OF EMPLOYMENT

The geographical decentralisation of office work is not new in the
UK and has been taking place for the past three decades, in part
encouraged for political reasons. For instance, office employment
in London as a proportion of office employment in the whole of the
South East region of England has been in decline since 1961. In
the years between 1963 and 1976 some 24,000 office jobs a year
left Central London and about 7,000 a year left London altogether,
mainly to other parts of the South East region. Of the moves
aided by the Location of Offices Bureau between 1963 and 1976,
four out of ten were from Central London to the South East outside
London; another three out of ten were to the Outer London
suburbs; and only four out of a hundred moves were to Inner
London [1].

A large proportion of these moves did not involve whole
organisations but only certain departments. In general, the
functions which were decentralised were the most highly automated,
involving work which followed standardised procedures and involved
relatively little face-to-face contact, according to studies
carried out by the Centre for Environmental Studies [2]. Data
processing and accounting departments were two of the functions
most likely to be relocated out of the city centre in this way.
Government Departments themselves contributed to this trend. The
Department of Health and Social Security, for instance, moved its

central computer department to the North East; while the computers for keeping track of company records and vehicle licenses were transferred to South Wales.

Political arguments played an important part in such moves. It was clearly advantageous wherever possible to shift jobs to areas of high unemployment and, as well as setting an example by relocating some of their own departments, successive governments developed a variety of incentives to encourage firms in the private sector to move to such areas. However this was by no means the only factor contributing to the trend towards decentralisation. For example, in a series of surveys carried out by the Greater London Council in 1976 and 1977, companies planning to move outside London were asked their reasons for relocating. By far the most common reasons were 'expansion' and 'modernisation', followed in only about a quarter of cases by 'consolidation' and 'labour shortages/recruitment problems'. Other factors were negligible [3].

The implication is that, although companies thought it advantageous to move where a suitably qualified labour force was likely to be easy to recruit, the most important factors were a shortage of accommodation of the right size and cost, and the fact that the company was 'modernising'. If we assume 'modernisation' to mean a combination of rationalisation and automation of work processes, then we are left with a picture which bears out the findings of the Centre for Environmental Studies researchers that highly automated processes are easier to decentralise than those which have not yet been extensively rationalised.

It can thus be argued that the introduction of computer technology, with its associated systematisation of work processes and reduction in the need for face-to-face communication, facilitated the trend for relocating employment out of city centre areas long before the additional potential for decentralisation created by the introduction of cheap microcomputers and telecommunications networks was recognised.

If information technology provided the means for decentralising some types of employment, it was economic realities which supplied the motive. City centre rents went up rapidly throughout the seventies, creating an urgent incentive to move to cheaper areas, particular during times when markets were stagnant and companies had to find ways to reduce their overhead costs to stay profitable. For many, decentralising some departments also meant access to a new, locally-based labour force, both cheaper and more stable than in the city centre.

The acceleration of this trend which took place during the late seventies and eighties has been little studied in Britain, although parallel developments in the United States have received some attention.

Three distinct types of relocation can be identified during this period. The first consists of moves away from metropolitan city centres to other regions. Into this category in the UK, as well as the transfers from London of large government departments already mentioned, come a number of moves made by companies in the finance sector, some down the 'M4 corridor' which runs from London to the west, to Bristol and South Wales, and some to East Anglia.

In a few cases back office functions involving extensive data entry have been moved to the North of England.

The second type of relocation has been out of city centres into the immediate commuter catchment area, a type of move which has been christened 'suburbanisation' where it has occurred in the United States. Mitchell Moss, who has studied the movement of office work away from Manhattan and into the New Jersey and Long Island suburbs of New York, attributes this trend partly to the rising cost of office space in the inner city and to employers' preferences for married white women clerical workers [4]. Similar conclusions were reached by Kristin Nelson who studied insurance companies in the San Francisco Bay Area and found that the transfer of work to the suburbs produced a workforce which was more likely to be white, college-educated, married, and more willing to work part-time and for lower wages and benefits than its inner-city counterpart. She concluded that 'The transfer of jobs from central city low-income, predominantly-minority female workforces to higher-income, predominantly-white suburban female workforces is not an unfortunate side effect of back office relocation necessitated by land cost considerations - it is one of the major reasons for back office relocation' [5].

Although there has been little systematic study of suburbanisation in the UK in recent years, there seems little doubt that it is taking place on a substantial scale in a number of industries which have previously been based in the city centre.

The third type of geographical decentralisation involves shifting work offshore, using satellite links for communication. Again, this is much better documented in the United States than elsewhere [6]. In general, only very routine non-time-sensitive data entry work is involved. Apart from some typesetting work, there is little evidence that this trend is significant in the UK, although Ireland is a site for some offshore processing for companies based in the USA [7].

Whether they are in a suburb, at the opposite end of the country or across the ocean, in a period when most office functions have been computerised to some extent, much of the communication which takes place between these decentralised offices and their control centres involves the transfer of electronically coded data. Does this mean that all the staff in these offices should be called teleworkers? To answer affirmatively would be to include a staggeringly high proportion of UK office workers within the category, many of them carrying out their work in traditional ways. To decide that only a proportion should be included creates a problem about how any sub-group should be delineated.

There are several possibilities. One could, for instance, relate it to the size of the organisation. At one end of the spectrum are huge unwieldy government offices like the Social Security computer department in Newcastle or the Vehicle Licensing one in Swansea. At the other are small suburban word processing pools serving city centre offices. But where along this spectrum should the line be drawn? And could it be anything but arbitrary?

Another possibility would be to include only those staff whose work involves the direct transmission of electronic data through

telecommunications networks. But this again seems likely to
produce arbitrary classifications. Should one distinguish, for
instance, between someone accessing a computer in the next room
and someone accessing a computer at the company's head office,
using identical methods? Is it realistic to draw a distinction
between data preparation operators involved in batch processing
whose work is transmitted to head office at the end of the day by
their supervisor and those involved in online processing who are
directly linked to the main computer? If a new system is
introduced into a local branch based on microcomputers, to
replace an earlier system using a head-office mainframe, do all
the workers using it suddenly cease to become teleworkers?

Pursuing questions like these makes it apparent that any
definition of telework which relies solely on geographical
decentralisation combined with the use of new information
technologies is likely to be unsatisfactory.

3.   EXTERNALISATION OF LABOUR

It was argued in the previous section that the introduction of
information technology plays a significant role in encouraging
organisations to decentralise. It can also contribute to the
redistribution of labour in situations where organisations are
already highly decentralised.

A useful concept for understanding this effect is that of the
externalisation of labour. When there is an electronic interface
between an information provider and the person who wishes to
access that information, it becomes possible to transfer labour
from one side of that interface to the other, generally from the
supply side to the demand side. This can take place within
organisations, between organisations or between organisations and
individual clients.

To illustrate how this effect takes place within organisations,
we can take the example of the insurance industry. Many UK
insurance companies have had a branch structure since their
inception. However the introduction of new technology has
brought changes in the division of labour between the high street
branches and head offices. Computerised databases have given
decentralised staff immediate access to information which
formerly had to be retrieved by telephone or letter from a
central office, allowing them to provide detailed comparisons
between different policies and instant quotations for their
clients. This has in turn enabled their employers to slim down
head office staffing levels, since fewer staff are now required
to process information requests from branches. Similar changes
have taken place within other financial institutions with branch
structures, and in retailing where sophisticated point-of-sale
terminals have made automatic ordering possible, along with the
automatic supply of information about stock levels to central
offices. Public sector organisations, such as Jobcentres or
social security offices also fall into this category when
computerised information systems are used.

A similar shift of labour has taken place between organisations.
The provision of videotex terminals in travel agencies, for
example, has enabled a great deal of information processing

relating to the availability and booking of holidays to be
transferred from the tour operators to high street travel agents.
The remote accessing of specialist databases has become common
for people in other occupations too, ranging from journalists to
mail order agents. In each case, it could be said that labour
formerly employed by the information provider has been
externalised.

The logical conclusion of this tendency, of course, is to
cut out the intermediate agency and for the ultimate consumers
themselves to access information directly. This has already
happened with automatic telling machines in banks and to a lesser
extent in such developments as home banking and home shopping and
various other personal uses of videotex services. However our
direct concern here is with paid work.

Once again, there are decisions to be made about which, if any,
of the workers involved in these externalised labour processes
are to be regarded as teleworkers. As with the recently
decentralised workers discussed in the previous section, if we
are to regard them all as teleworkers we will end up with an
enormous total, comprising all employees of organisations with
branch structures which make use of information technology for
the internal communication of information and all employees of
organisations which regularly access databases provided by third
parties. Together these make up a substantial proportion of all
service sector employment. To define them all as teleworkers
would produce a category so broad as to be meaningless.

However any attempt to separate sub-sections of this group for
classification as teleworkers is again fraught with difficulties.
To use the insurance industry once again as an example, there is
a continuum running between the individual insurance broker or
'financial adviser', paid by commission, operating from a home
base and using a videotex-based network as an essential tool of
the trade, and the office of a large insurance broker: the
difference is only one of the scale; the relationship with the
information provider and the technology used is likely to be
essentially the same. The difference between the work carried
out in the office of the insurance broker and that done in the
branch office of a company which is itself an insurance provider
will again be very similar. To confuse things further, many of
the jobs carried out in the branch office of the insurance
company may be identical to others carried out in the same
company's head office.

Examples drawn from other industries lead us to parallel
conclusions. We are therefore left without any obvious point at
which to place the dividing line which separates teleworkers from
their fellow workers. Should it depend on their location of
work? On the technology they employ to carry it out? Or on
whether the network to which they are linked is provided by their
employer or by some other, external organisation? Or should we
simply decide that this factor is entirely irrelevant to the
definition of telework?

Whatever the answer to these questions, it seems that we will
have to look to other variables to find a conclusive definition.

4. CONTRACTUAL CHANGES

A third set of parameters for a potential definition of telework is suggested by the changes which are currently taking place in contractual relationships between employers and workers. Here the trends already mentioned play a part, although there are several other factors at work too.

One facet of this is the growth in sub-contracting which has become evident in the UK, as in most other industrialised countries, during recent years. The introduction of information technology has contributed to this growth by making it possible to hive off many functions which were previously seen as integral to the internal functioning of organisations. Office automation facilitates the 'unbundling' or disaggregation of organisational structures by standardising processes, formalising decision-making structures and increasing the potential for quantifying and monitoring the performance of individual parts of an organisation [8]. The result has been a vertical disintegration of organisations and an increase in the subcontracting of a wide range of services, often to companies started up by ex-employees of the contracting organisation.

As with the trend towards decentralisation, the technology should be regarded as a means, not as a motive for the implementation of change. As Swasti Mitter has shown, the decision to subcontract is generally taken on economic grounds, as a way of cutting down on permanent overhead costs, and to provide maximum flexibility in uncertain times [9].

Another contributory factor to the growth of subcontracting is the very tendency to geographical decentralisation described earlier. This can be illustrated by the example of the publishing industry. Until the end of the 1960s, book and magazine publishing in the UK was heavily concentrated in the Central London area. However during the 1970s and early 1980s there was a massive exodus to the outer suburbs, partly to accommodate the much greater size of individual organisations caused by a spate of mergers, and partly to reduce costs at a time when some companies were in financial difficulties. The development of computerised warehouses and the need to integrate these with other functions also played a part. Most companies kept a slimmed-down head office in Central London. An advantage of the move to the companies was that they could now attract and retain a well-educated and relatively undemanding corps of clerical workers. However the recruitment and retention of creative and professional staff proved more difficult, despite an oversupply in the industry.

The typical editorial worker of the 60s and early 70s was fairly young and ambitious. Career development was achieved by hopping from one company to another - staying more than two years in a job was considered to be stagnation, unless it was a senior post. Staff were regularly poached from one company to another and so were ideas. No only did most publishing workers like living in the city centre and mixing socially with colleagues from other companies; this participation in a sub-culture of media workers was also of benefit to the employers, since it contributed to the creative success of their enterprises - contacts with new authors were made at social gatherings; articles were commissioned at dinner parties and so on. A critical mass of publishing companies in the city seemed necessary to permit this subculture to flourish and also to support the existence of specialist suppliers to the industry.

This group of workers was on the whole not well pleased by the shift out of Central London. Many continued to live there and became 'reverse commuters' when their employers moved to the suburbs, but kept an eye out for alternative jobs. Others moved near to their work and concentrated on making a career within their own company, breaking with the tradition of progressing by lateral hops between organisations. However the most noticeable new development during this period was the very large number of employees who went freelance, either on an individual basis or by forming the subcontracting companies which are known in the industry as 'packaging out houses'.

Of course the trend towards decentralisation was not the only cause of this phenomenon. It could also be attributed to a response by the employers to the unionisation of the publishing industry which took place during the same period; and it could also be seen simply as a part of the general trend towards subcontracting as many services as possible and, in the Japanese phrase, hiring workers 'just in time' instead of 'just in case'.

Nevertheless, this development can be seen as a solution to the dilemma of retaining a critical mass of creative and professional workers and supporting services in the cultural centre while permitting employers to take advantage of the larger, cheaper premises and more compliant clerical and warehouse staff available in the outer suburbs.

It is possible that other industries which employ specialist professional staff or require a major creative input, such as software design, may exhibit a similar pattern. Certainly the software industry is another which is characterised by a high degree of fragmentation, with many small companies and individual freelances.

The decline in the proportion of the workforce which is full-time and self-employed cannot be attributed entirely to increases in subcontracting and the growth of self-employment. There has also been a growth in temporary contracts and part-time working, which carries less employment protection and fewer employee benefits than full-time employment. The UK has not yet seen the introduction of 'minimum-maximum' contracts on the Dutch model, however.

Some aspects of these trends can be witnessed in the employment statistics. Excluding craft and manual workers, the number of workers registered as self-employed, for instance, grew from approximately 1m to 1.68m between 1981 and 1984. Part-time workers covered by the annual New Earnings Survey rose from 2.7m (compared with 14m full-time workers) in 1980 to 2.9m (compared with 13m full-time workers) in 1985. This figure is an underestimate of the real number of part-time workers since the survey only covers those earning enough to be paying national insurance contributions. Many part-time workers are known to earn well below this level.

Data from the bi-annual Labour Force Survey show that casual, temporary and short-term contract workers rose as a proportion of the total workforce from 2.7% in 1981 to 5.5% in 1983 and 6% in 1985. These percentages are also likely to represent under-estimates since they do not allow for employment in the underground economy, most of which is casual in nature [10].

Other features of this tendency, such as the growth of subcontracting, are harder to quantify. However the most authoritative recent survey of the data on small firms, carried out by James Curran for the Small Business Research Trust, concludes that 'for a wide variety of small enterprise, there has been a remarkable increase since 1970 so that the overall total and proportion of small scale activity in the economy has increased' [11]. The statistics do not, unfortunately, give us any indication of the relationship between those small enterprises and larger ones. Are they carrying out work previously undertaken within larger organisations? Or are they engaged in new activities? Commentators like Philip Mattera are convinced that increasing subcontracting is a major cause of the growth in small enterprises both in the official and in the underground economy [12]. However no methodology has yet been devised for substantiating this claim other than case-studies of particular firms and industries which may not be typical of the economy as a whole.

Although some of the workers normally described as teleworkers in the literature on the subject have permanent full-time employment contracts, these are comparatively rare. Subcontracting, self-employment and temporary contracts all feature prominently, and permanent employees are frequently part-time. The search for flexibility in contractual relationships with employees has undoubtedly contributed to the development of what is commonly thought of as telework.

But whether work is contracted to a subcontracting company, to an individual freelance or a temporary contract worker, we are still left with the problem of which types of subcontracted work to classify as telework.

Even if we restrict the range of possible activities to office services, the dilemma does not disappear. A typical office might subcontract such services as graphic design, typesetting, advertising, accounting, legal advice, data processing and overflow copy typing all of which involve the processing of information to some extent. If they are not all to be classed as telework, how should distinctions be made? Should the designation be restricted only to those using computers? If so, at what level of technology usage is a worker reclassified? Let us say that a self-employed copy typist trades in her ageing electric typewriter for an electronic typewriter with a memory. Is she now a teleworker, or must she wait until this has in turn given way to a word processor? If the graphic design studio acquires a microcomputer for doing the accounts, are the designers transformed overnight into teleworkers?

An alternative demarcation might be drawn between those using a direct telecommunications link with the contractor for the delivery of information. However this too might lead to arbitrary distinctions between groups of workers whose location, method of work and relationship with the employer do not differ significantly in other respects. For instance, our copy typist, equipped with her new word processor, might find it more convenient to send the floppy disc containing her completed work to the contractor by messenger rather than using her modem. Does such a small difference justify a separate classification? Similarly, one could argue that the nature of a graphic designer's work is not fundamentally altered because the copy and rough sketches for a rush job happen to be delivered via a fax terminal, instead of through the post.

A distinction based on the size of the subcontracting enterprise is also likely to produce arbitrary results. There is rarely any real difference in the work carried out by a single freelance, two who have got together to form a partnership, a small firm employing perhaps four or five, or a larger company. Such differences as do exist, for instance cost differences caused by the higher overheads of organisations which are not based in people's homes, or the additional reliability offered by an organisation, as against an isolated individual, in case of sickness, have no connection with the contractual relationship, the technology employed, or the nature of the work carried out.

It appears then that even a study of contractual relationships, although it may spread useful light on some aspects of telework, fails to provide us with the raw material for a definition of telework which can form a framework for empirical research.

5. GROWTH OF HOME-BASED WORKING

Another possibility, although it is one which is frequently rejected by researchers into the subject, is to regard telework as a sub-category of homeworking. (The reason for this rejection, one might guess, sometimes has more to do with dodging the political criticism often directed at advocates of homeworking than with producing a coherent framework for analysis.)

This approach would at least appear to have the advantage of making a clear distinction between telework and office-based work of the traditional type, which, as we have seen, becomes blurred using definitions based on geographical decentralisation, externalisation or contractual relationships.

But even here it soon becomes apparent that there is no hard and fast boundary to be drawn. One important variable is the amount of time spent working at home. It is very common for workers who are normally office-based to take work home occasionally, perhaps because it requires concentration to complete, because it needs to be finished urgently or because the worker is ill or temporarily disabled. it is also common for workers who are primarily based at home to visit the office of their clients or employers and sometimes to work there for long periods. In between, there are a number of flexible arrangements which do not fall easily into either category. This difficulty is not insoluble in practice. It would be relatively easy to design a survey which, for instance, defined as homeworkers those who had spent more than 50% of their working time at home over a given period.

However that is not the only difficulty. It is also important to distinguish between people who work 'at home' and those who work 'from home'. Here too there is a grey area in the middle, since there are some workers who work partly at home and partly elsewhere.

Many current definitions of the term 'homeworker', exclude those working 'from' home, as well as professional and artistic workers. For instance the Homeworkers (Protection) Bill, unsuccesfully introduced to parliament in 1979, defined a homeworker as

> 'An individual who contracts with a person not being a professional client of his or hers for the purpose of that person's business for the execution of any work (other than the production or creation of any literary, dramatic, artistic or musical work) to be done in domestic premises not under the control of the management of the person with whom he contracts, and who does not make use of the services of more than two individuals in the carrying out of that work'

Adopting this definition might exclude a large proportion of the workers currently described as teleworkers, including freelance computer programmers, systems analysts, consultants of various sorts and some journalists, not only because of the professional nature of their work but also because it takes them out and about onto the premises of clients or other third parties, such as information suppliers.

The most comprehensive survey of homeworking in the UK carried out to date, by Catherine Hakim at the Department of Employment, took a much broader definition, including all people working at or from home, except for construction, transport, haulage and family workers. It took the form of supplementary interviews to the 1981 Labour Force Survey and produced a grand total of 660,000 homeworkers [13]. Of these, however, only 251,000 were working 'at' home of whom 72,000 were engaged in manufacturing work. The accuracy of this survey's findings have been called into question by Swasti Mitter who, after carrying out a detailed study of employment and output statistics in the UK clothing industry, concluded that the number of homeworkers manufacturing women's, wear children's wear and lingerie alone increased by 22,000 between 1978 and 1982 [14]. This suggests that the Labour Force Survey findings severely under-estimate the numbers of homeworkers in the UK, a view which is shared by Helena Pugh who compared alternative methods of estimating the extent of homeworking in a study based at the Social Statistics Research Unit at the City University [15]. Whatever the accuracy of the survey's findings, it clearly encompasses a much wider spectrum of workers than is normally included in discussions of telework.

Using other sources, such as the Census of Population, it becomes clear that even if we exclude manufacturing workers, workers 'at' home include groups such as farmers, hoteliers, publicans and shop-keepers who live and work on the same premises as well as a variety of artistic and craft workers. Similarly, excluding lorry drivers, taxi drivers, building workers and other clearly defined groups from the workers 'from' home still leaves us with a motley collection of sales representatives, service personnel, agents and entrepreneurs across a broad range of industries.

Again, any attempt to delineate sub-groups to be categorised as teleworkers is fraught with difficulties.

How, for instance, should one categorise farmers, TV repair technicians, plumbers, architects or any other group of self-employed, home-based workers making incidental use of computers for accounting, word processing or other administrative functions, and can they be said to differ in any significant way from others sharing the same occupations who use more traditional methods to administer their work, or who farm these functions out to a sub-contractor?

To take another example, what should be the designation of a married woman who is employed by her husband to keep the books and carry out secretarial duties for his home-based small business? She is, after all, a homeworker, but not working at a distance from her employer. If it is decided not to call her a teleworker, what happens if his business expands and he decides to work from an office or showroom elsewhere, although she continues to work from home - does she then become a teleworker?

Such examples could be multiplied endlessly. They serve merely to point out that the simple fact of working from a home base is not a good enough basis from which to construct a definition of telework which is likely to be useful for analytical purposes.

6. CHANGES IN JOB DESIGN

So far we have looked at factors relating to the technology used, the location of work and the relationship between the worker and the employer as possible aids to the construction of a definition of telework. None of these, however, has been satisfactory. Whether the work is decentralised or externalised, whether it is carried out in a head office, a branch office, the office of a subcontracting organisation or in the home; whether the worker is an independent contractor, an employee or a temporary contract worker; whether or not the work involves the use of a computer or of an on-line telecommunications link: none of these has proved sufficiently exclusive to provide a workable basis for a definition which might enable the researcher to quantify and analyse telework.

One further possibility is left to explore: whether there might be factors intrinsic to the design of some jobs which enable them to be defined as telework.

Some of the changes in job design resulting from the introduction of information technology have already been mentioned in the context of the contribution they have made to the decentralisation and externalisation of work, and to the growth of subcontracting. There are, in addition, other ways in which information technology has changed the design of jobs which also have a bearing on our subject.

There is, of course, an enormous literature on the specific ways in which various types of job have changed as a result of technological change, which it would not be possible to summarise here. Instead, it is proposed to list only those general factors which seem to relate directly to the development of telework as it seems generally to be understood.

Firstly, the introduction of information technology has led to the routinisation and standardisation of some types of tasks which has in turn resulted in a reduction of the need for face-to-face contact between workers. This has lessened the need for them to be situated close together and has contributed to the geographical decentralisation of some highly automated back-office functions, such as database management and accounting.

This standardisation has also made it easier to create a clear division of labour between functions which were previously integrated quite closely together and to quantify their performance. This has made it possible to restructure organisations so that different departments can become separate profit centres or be subcontracted in their entirety.

Another aspect of this new ease of quantification has been to make it very much simpler to monitor workers' performance accurately. This may mean direct machine monitoring, for instance in the measurement of key depressions, error rates, terminal link-up time or other variables; by monitoring output, for instance the number of reports completed, records dealt with or clients serviced; or by setting targets and measuring how well they are met. Such forms of monitoring, which do not depend on the physical presence of a supervisor, make it very much easier to employ workers at a distance from the control centre of an organisation.

Another effect of the introduction of information technology, already mentioned in the discussion of the externalisation of labour, can be to change the division of labour between a central office and its local branches or between a central supplier of information and its clients, changing the nature of the tasks carried out at the core and the periphery. This facilitation of the remote accessing of information can also be seen as a making a contribution to the decentralisaiton of work.

In some industries, automation has been accompanied by major changes in the skill requirements for certain jobs. Although at first sight this would seem to have little connection with changing their location, in practice it does appear to have had precisely this effect. The reason for this is that skill changes are often accompanied by changes in the sexual division of labour, opening up traditionally male jobs for women. And it is women, of course, who are most likely to be in the market for home-based, casual employment. One example of this is in the printing industry, where typesetting, traditionally a 'dirty', male job, based in a print-shop and requiring apprentice training, now merely requires the skills of a copy-typist capable of using a QWERTY keyboard. As a result, typesetting is increasingly being carried out be women homeworkers, previously employed in offices as typists or word processor operators, who are now tied to their homes by the need to care for young children. Anecdotal evidence suggests that this development is widespread in the UK, although it has not been as well documented there as it has in Germany, in the work of Monika Goldmann and Godrun Richter, at the Dortmund Sozialforschunsstelle [16].

If we were to take any of these changes in job design in isolation and attempt to use it as a defining parameter in new definition of

telework, it is likely that we would end up in similar
difficulties as with the earlier attempts to define the term using
other factors. Not all jobs which are monitored by machine are
likely to be telework as it is usually perceived; neither does a
skill change involving the feminisation of previously masculine
work necessarily imply a change in location. However it is
possible that by putting together several of these
characteristics, perhaps in combination with some of the other
factors looked at in this presentation, we might come up with a
definition which, albeit unwieldy, comes somewhere near defining a
category of work which it is possible to pin down and examine.

Before this hypothesis can be tested, a great deal of empirical
work will be necessary, analysing the design of jobs which have
already been relocated following the latest wave of technological
innovation and examining the ways in which they differ, on the one
hand, from jobs still performed in their traditional locations,
and on the other from the types of work which have traditionally
been carried out from the home, or on a freelance, temporary or
sub-contract basis.

7. CONCLUSIONS

These many difficulties facing the researcher attempting to
produce a clear definition of telework prompt the question, why
bother? If a concept is as nebulous as this, then wouldn't it be
better to abandon it and develop a different framework for
analysis, more amenable to precise categorisation? Perhaps,
instead of focussing on telework, we should be concentrating on a
study of the changing geographical organisation of work, new forms
of contractual relationship between employers and workers,
alterations in the design of jobs or some other tendency.

I believe that this would be a mistake. Despite its lack of
precision, the word telework has acquired a potent symbolic value.
Quite independent of the extent to which it describes a measurable
reality, the idea of the teleworker has become a representation of
what the future of work might be. Aware that major upheavals are
taking place in the organisation of work, though not quite able to
pinpoint what these are, many people have seized on this image,
and projected onto it many of their hopes and fears about the
future. Many of the employment trends discussed in the earlier
sections of this paper have been experienced in ways too poorly
articulated to be easily discussed. Nevertheless, they have
aroused considerable anxiety in some, and excitement in others.
The notion of telework provides a concrete focus for these
feelings.

It is no accident that the issue of electronic homeworking has
become so politically charged, and the debates surrounding it so
polarised. It can be seen as the logical conclusion, the most
extreme form, of many of the tendencies discussed above. The
image of the remote worker, connected to the world of work only by
a sort of electronic umbilical cord, has become a crystallisation
of the dreams and nightmares of many people directly caught up in
the currents of change swirling through our industrial structures.
For some it is a symbol of liberation from the daily grind of
living in a suburb, commuting into the city, working from 9 to 5
and wearing stuffy suits, a chance to get to know one's family and

pursue new leisure interests. For others, it has become a symbol of isolation, exploitation and the end of any possibility of temporary escape from the drudgery of housework and childcare.

Both of these attitudes are rooted in direct experience. For many men, and some women, the home *is* a place of rest and relaxation, an escape from the stressful world of the office. For many women, and some men, it *is* a place where stress is generated and chores seem unending, a place from which to escape whenever possible into the relatively pleasant social environment of the office. Attitudes to being at home are not, of course, the only variable. For instance, peoples' willingness to work in solitude also varies enormously according to how creative or interesting or well rewarded their job is. While these differing experiences remain, so will the controversy about telework.

It thus becomes a matter of direct public concern to measure the extent to which this highly charged image is in fact becoming a representation of reality. To do so, I have argued here, is no easy matter, since it involves the analysis and quantification not of one tendency, but of several. Nevertheless, it seems necessary to make the attempt. The first step will be to understand the complex interplay between these various tendencies and develop methodologies for quantifying them. Only then, I believe, will we be in a position to discuss telework with any degree of confidence, and to frame policies to encourage its positive aspects, or limit its negative ones.

c   Ursula Huws, 1987

8.   REFERENCES

[1]   Barker, J., and Huws, U., Office Work and Information Technologies, Strategy document no 10 (Economic Policy Group, Greater London Council, London, July 1983)

[2]   Marquand, J., The Service Sector and Regional Policy in the UK, Research Series No 29 (Centre for Environmental Studies, London, July 1979) and Locational Analysis of the UK Service Sector, Occasional Paper No 9 (Centre for Environmental Studies, London, November 1979)

[3]   Haydon, F.W., Factors Influencing the Location of Industry, Research Memorandum 528 (Department of Planning and Transportation, Greater London Council, London, 1978)

[4]   Moss, Mitchell L., New Telecommunications and Technologies and Regional Development (New York University, New York, 1984)

[5]   Nelson, K., Back Offices and Female Labor Markets: Office Suburbanisation in the San Francisco Bay Area (University of California Dissertation, Berkeley, 1983)

[6]   US Congress, Office of Technology Assessment, Automation of America's Offices (US Government Printing office OTA-CIT-287, Washington, December, 1985)

[7] The Instant Off-shore Office (Business Week, March 15, 1982 p. 136)

[8] Brusco S., Labour Market Structure, Company Policies and Technological Progress, in Diettrich, O., and Morley, J., (eds), Relations between Technology, Capital and Labour (EEC, Brussels 1981)

[9] Mitter, S., Common Fate, Common Bond: Women in the Global Economy (Pluto Press, 1986)

[10] Department of Employment data

[11] Curran, J., Bolton Fifteen Years On: A Review and Analysis of Small Business Research in Britain 1971-1986 (Small Business Research Trust, London, September 1986)

[12] Mattera, P., Off the Books: the Rise of the Underground Economy (London, Pluto, 1985)

[13] Hakim, C., Homework and Outwork: National Estimates from Two Surveys, Employment Gazette (January, 1984)

[14] Mitter, S., Industrial Restructuring and Manufacturing Homework, Capital and Class No 27 (Winter 1986)

[15] Pugh, H.S., Estimating the Extent of Homeworking, Working Paper No 15 (Social Statistics Research Unit, the City University, 1984)

[16] Goldmann, M., and Richter, G., with the collaboration of Wasserman, W., Business Interests in Flexibility and the Origin of Home-based Teleworkplaces for Women: Empirical Examples from the Printing Industry, in Gehrmann, F., (ed), Neue Informations und Komminkationstechniken, Ansatze zur Gesellschaftsbezongenen Technologieberichterstattung (Campus-Verlag, Frankfurt am Main/New York, 1987)

ORGANIZATIONAL BARRIERS TO TELEWORK

Margrethe H. Olson

Associate Professor and Director, Center for Research on Information Systems, Graduate School of Business Administration, New York University, 90 Trinity Place, New York, NY, 10006, U.S.A.

This paper discusses telework as a work option for full-time organizational employees. Based on research of the author and others, organizational barriers to telework are identified. The author concludes that these barriers are significant, and are related primarily to organizational culture and management style rather than information technology.

1. INTRODUCTION

This paper addresses the current trends in telework in the United States, particularly trends related to organizational forms of telework. It begins with a definition of telework as used within this context, and discusses the categories of work and workers for which the term is commonly used. Reasons for employee and organizational interest in telework as an employee work option are identified. The actual current trends to telework in the U.S., and related research, are discussed. Predictions regarding future developments in technical, economic, organizational, legal, and social/psychological domains with respect to telework are given. The paper concludes with some future organizational scenarios regarding work organization and the organization -- employee relationship.

2. TYPES OF TELEWORK

A significant problem with the research and writing on telework to date has been the use of the term to refer to different phenomena. It is therefore important to begin with specific definitions and clarifications of the various forms of telework.

2.1. Definition of telework

In general, the term telework is used to refer to organizational work performed outside of the normal organizational confines of space and time, augmented by computer and communications technology. The work is not necessarily performed in the home. Because of the role of technology, telework is generally confined to work that would otherwise be performed in an office, as opposed to industrial work. Information technology potentially removes certain constraints on where and when office work can be performed.

This section identifies the different types of work that fall under the general category of telework and defines the specific domain which is addressed by this paper. Much of the controversy about telework, particularly work in the home, is exacerbated by a generally confused notion of who works at home, under what circumstances, and for what reasons. The primary categories of importance are where, when, and under what employment status a person performs telework.

2.2. Where is Telework Performed?

While most of the interest in telework implies that the work is done in the home, this is not strictly the only telework option available. In the broadest sense, telework includes any situation where the employee is physically separate from the employer. This could include physical decentralization of functional units as well as "off-shore" work. Both of these phenomena have been common with industrial work and are now becoming more common with office work, as in back-office decentralization and off-shore data entry. Two more innovative options which take advantage of the potential of information technology, as described in [21], are satellite work centers and neighborhood work centers.

When an organization locates a regional office based on the residential location of its employees, it may be referred to as a satellite work center. The difference between this and decentralization of a functional unit is that any employee who lives near it may work at the satellite work center rather than the central office, regardless of organizational function. This assumes, of course, that the necessary resources to support the employee's work (primarily computer and communications equipment) are available at the satellite work enter. The result of such an arrangement is that while employees may work regular hours and have their own office space, they are performing unrelated functions.

Although Nilles and his colleagues suggested satellite work centers as a way to save energy in 1976, few companies have actually implemented the concept. Control Data Corporation and Southern New England Telephone have set up formal pilots. Other companies have set up satellite locations to perform single functions; an example is Travelers Insurance, which set up "remote programming facilities" in suburban locations and hired local residents to staff them. Each facility employs about 100 people, primarily women, as programmers. When a single function is performed and supervision is on-site, the difference between satellite work centers and back-office decentralization becomes moot.

A neighborhood work center is a shared office facility, where employees from many different organizations as well as self-employed workers share resources in a common facility. It may be equipped with teleconferencing facilities, clerical support, and even day care facilities as well as computer and communications equipment. Workers go to the neighborhood work center nearest their home to perform their duties. In the U.S., there have been no formal experiments with neighborhood work centers to date.

Another, more informal category of telework is occasional work away from the office, such as in a hotel room or in transit (e.g., airplane, train, automobile). There is a considerable amount of technology which has been applied to this type of telework: video conferencing, lap-top computers, and automobile telephones are examples. The emphasis is primarily in providing the ability for a travelling employee to "keep in touch" with the office.

The most common telework location is of course the home. Unless explicitly stated, in the remainder of this report it will be assumed that telework refers to work performed in the home. The categories below, when and under what employee status, refer only to telework performed in the home.

2.3. When is Telework Performed?

While most popular news magazine stories on work at home imply that this is the exclusive work location, this is generally not the case. Most people who claim to telework are regular organizational employees who spend substantial amounts of time at home doing job-related tasks on computers or terminals in addition to their regular eight-hour work day at the office. The term electronic briefcase sometimes refers to this phenomenon; the technology facilitates extending the work day into the home. Employees who formerly worked overtime in the workplace because of the need to access computers or physical materials can now accomplish this work at home instead. Some companies have been quick to perceive the value of making the necessary equipment easily accessible; the increased volume of work output easily justifies the additional cost of equipment. The acceptability of such work habits to the employee are typical of the "computer culture" in which the hours spent at the computer have little relation to scheduled work hours [29]. Many professionals and managers, if their jobs have sufficient autonomy and status, work at home occasionally to escape the interruptions of the office or to finish a report for a critical deadline. In some organizations with a large proportion of professional employees, staying at home on occasion (e.g., one or two days a month) is becoming a relatively common phenomenon.

Another type of informal arrangement is the special case of allowing an employee to work at home several days a week for a temporary period and providing the necessary equipment in the home. This type of informal arrangement is becoming more frequent in the prototypical case of an employee on maternity leave who is highly valued and may otherwise decide to terminate. Often such arrangements are left informal and kept quiet because they run counter to human resources policy. It is not uncommon for the human resources department to prohibit employees from working at home for a variety of reasons including legal bans (which vary by state as well as industry), insurance liabilities, challenges to Workmen's Compensation, etc. These issues are discussed further in a later section under employee status.

Formal or permanent telework arrangements include all those in which a person works at home as a regular substitute for work at a separate work place, either part time (e.g., two days a week) or full time. A formal arrangement of this type is common for writers, consultants, professors, etc. This category of course also includes full-time homemakers as well as artists and crafts persons. It is also important to note that many workers who fall into this category really work "out of their homes" as opposed to in them. For instance, consultants and salespersons may have their only office in their homes, but they spend the work day on the road calling on prospects, or at a client's site.

A formal arrangement to substitute work at home for work in the office from one to five days a week, agreed upon between the organization and a full-time employee, falls into this category.

Although popular news stories tend to imply this category of arrangement, their examples tend to be of the other categories. This paper will demonstrate that this type of arrangement is in fact today a relatively rare phenomenon in the U.S.

2.4. What is the Worker's Employment Status?

People work at home under a variety of conditions of employment. There has been considerable confusion caused by the tendency to generalize across different conditions. A greater problem has been the tendency to attribute abuses (or potential abuses) of workers to the fact that they are at home rather than to the conditions of their employment.

One class of employment is the full-time, salaried employee who receives full salary and benefits while working at home either part time or full time. Commonly, when an organization sets up a pilot telework program, the employee's status remains unchanged for the duration of the pilot. Thus changes (e.g., in performance) can in large part be attributed to the work location (e.g., fewer distractions, lack of co-worker interaction), rather than changes in employee status or compensation.

Arrangements with non-exempt (i.e., from overtime compensation) employees involve either hourly pay or piece rates. Presumably, when an employee is on-site the company primarily controls the hours worked; since this is not the case when the employee is at home, companies often prefer a piece rate. As an example, Blue Cross and Blue Shield of South Carolina instituted a work-at-home arrangement based on piece rates while comparable work being performed in the office remained on an hourly wage. In addition, the employees working at home do not receive benefits. In such a case it is not valid to compare performance between the two groups. While union opposition to work at home is well known, much of the opposition is based on the potential for companies to exploit employees through low piece rates combined with high quotas. While the potential for abuse is an important issue and should not be dismissed, it is not necessarily an issue of work at home. Companies can set fair piece rates and adequate protections for employees working at home. On the other hand, companies can set unfair piece rates and quotas for employees working on site as well as at home.

Recent congressional testimony on work at home focuses on the issue of employment status and raises important questions [30]. A primary issue is whether the worker is actually on independent contractor or employee status, and what rights the worker has with respect to each.

The vast majority of people who earn some type of income working at home are, either formally or informally, independent contractors. Recent books on work at home refer to a presumably new category of workers who prefer autonomy to security and a steady income and are setting up their own businesses at home [1,7,11]. These books [e.g., 7] are primarily about setting up one's own business, which happens to be in the home. As will be discussed in more detail in a later section of this paper, interest in this work option is primarily motivated by personal needs for flexibility in order to accommodate nonwork responsibilities (e.g., family) as well as a personal desire for autonomy.

2.5. Telecommuting

None of the categories discussed above explicitly requires the use of computer and communications technology in performing the work. If technology is used, the work can also be categorized as telecommuting. A formal definition of telecommuting is: the use of computer and communications technology to transport work to the worker as a substitute for physical transportation of the worker to the location of the work. Throughout this report, the terms "telework" and "telecommuting" will be used interchangeably and unless otherwise specified, will refer to this definition.

3. CURRENT INTEREST IN TELEWORK IN THE UNITED STATES

If work at home is not a new phenomenon, why is it creating so much interest (and controversy) today? In this section, some of the social and economic forces affecting individuals and organizations in the U.S. are discussed. The role of information technology with respect to work at home is put into perspective.

3.1. Why Do People Want to Work at Home?

In preliminary surveys and interviews with people who work at home [6,24,25], I identified several recurrent issues with respect to people's interest in work at home. They are the following:

* Need for flexibility. With over seventy percent of women, and over fifty percent of mothers of small children, holding permanent jobs, the amount of conflict between work and nonwork demands for both men and women has increased substantially. Flex-time programs are very limited in scope and do not begin to address the real need. Workers search for any kind of work situation that gives them back control over their own work schedule, and work at home appears to provide that control.

* Desire for autonomy. There is some indication of an increase in the number of people who choose autonomy over job security in their work lives. For the most part, the desire for autonomy is addressed by setting up one's own business, and the home is a logical place to start because of cost considerations. There is some question as to whether the number of self-employed professionals is really increasing; this will be discussed in the next section in the review of research on census data. Furthermore, in many cases what is described as a desire for autonomy is really dominated by needs for flexibility as described above.

* Commuting hassles. The tolls of commuting to and from work, on stress and physical health as well as time and the pocketbook, have not been adequately studied. For many, the value of even one day a week at home is primarily felt in terms of not commuting; they feel much better and add as many as four hours to their productive day. The effect of commuting stress on productivity has not been recognized; it is possible that for many, at least the first half hour of the day is spent simply recovering from getting to work.

* Limited alternative work options. For many, a job outside the home is simply not within reach. A stereotypical case is

a woman with small children and few or no skills in demand, to whom the only jobs available entail the costs of commuting, clothing, and child care (if it is even available). To these people work at home may be the only option. When asked, people who work at home under these conditions are very happy with their work arrangements [3]; the income is badly needed and the alternative is not working at all.

* Lifestyle demands. For a small number of people whose skills are in demand, work at home is a convenience and a privilege. It may be that they choose to live at a distance that precludes commuting, and the company is willing to permit them to work at home because of their valued skills. Others see the benefit because of hobbies and recreation; they can play ball with their kids when they come home from school, or ski in the middle of the week when there are no crowds. A disproportionate amount of attention has been given to people in this situation, such as wealthy stockbrokers and specialized computer "hackers".

3.2. Why Do Organizations Want People to Work at Home?

An important question which is not obvious is why organizations would be interested in telework as an employee work option. In my own preliminary surveys [6,23], I identified some consistent themes.

Organizational interest in telework experiments is spurred primarily by short-term needs, and the most pressing need is to attract and/or retain qualified employees. Shortages of qualified employees are particularly acute in the data processing profession, a primary reason why many pilots originate in data processing departments. Sometimes a need is spurred by an immediate situation and a need to respond; one pilot I studied began because the department was relocating and management was looking for ways to retain key employees who would otherwise leave because of long commute times. More often, the pilot is used to demonstrate that telework is feasible; the next stage is presumably to hire new employees who are highly qualified but would be unavailable to the firm without the arrangement.

A second short-run interest on the part of some organizations is productivity improvement. Although most managers are only concerned that productivity does not decrease while the employee is at home, others recognize that significant productivity gains are feasible. There are several possible reasons: The employee may work longer hours, but more likely the employee only counts the time he or she is working. Breaks to do the dishes do not count as work time, whereas in the office informal breaks are part of the work day. Another possibility, particularly with programmers, is the opportunity to solve a problem (e.g., fix a programming bug) when the person thinks of it. For instance, there are stories of programmers thinking of a solution in the middle of the night; if the necessary equipment is in the home, the person gets up and tries out the solution instead of waiting until morning and possibly forgetting it.

A third reason for organizational interest in telework is simply faddism. With the press focusing on telework, a company may receive favorable publicity for its "enlightened" work style. One company hired twelve physically home-bound disabled at considerable expense and benefited from the publicity. Others are concerned that if the option does prove to be widespread, they need to be ready. In general, it is viewed by personnel departments as another interesting work option to be studied.

Management often describes long-term scenarios when discussing telework. In general, significant savings in indirect costs are envisioned: space, cafeteria service, parking, etc. Furthermore, there is often an implicit association made between work at home and contract work, so that savings from moving employees to contract status are envisioned. The basic premise is: if telework is feasible for large numbers of workers, then the organization can enjoy significant savings from reducing many kinds of overhead, including but not limited to employee benefits associated with supporting full-time employees at a work site. It is clear that such savings would only be realized if a significant percent of the employee population were shifted permanently into their homes.

3.3. The Role of Information Technology

The role of information (i.e., computer and communications) technology in the phenomenon of telecommuting is significant. However, as should be clear from the preceding discussion (which said very little about technology), it is not the driving force. The basic premise of this paper is that information technology facilitates new forms of work organization, of which telework is only one example. Organizational culture and individual needs play a much more significant role in determining what new forms will be adopted.

In the following discussion, it should be kept in mind that the primary focus here is on office work. The primary tools required to perform office work are changing from paper, pens, telephones, calculators, and typewriters to personal computers. If a person's primary tools are personal computers and telephones, the person can use that equipment at home with relative ease. More important, however, are the subtle changes in interdependence. If I as a professional writer no longer need the services of a professional secretary to format my manuscripts, physical proximity to that secretary is no longer a critical need in my work. If I can send a completed document over telephone lines to be printed out and distributed by that same secretary, both parties can fulfill their functions more efficiently and effectively than before regardless of my location when I perform the work.

Many tools are now available which make work more portable: satellite and automobile telephones, lap-top computers, remote-access answering devices are a few current, affordable examples. To a person starting a new home-based business, a basic answering machine is probably more important than a personal computer. Other tools, primarily communications support tools, reduce the constraints on time at work. Electronic mail, unlike the telphone, is generally person-bound rather than location-bound so the location as well as the time constraint is removed.

It is important to note that the level of penetration of information technology into basic office work which would really relax the constraints on work in space and time have not been achieved to date. While personal computers are proliferating rapidly, they are far from becoming a basic office support tool of the nature of a typewriter or telephone. Furthermore, substitution of electronic for other forms of communication is a key requirement of widespread telework; a person working remotely must be able to keep in touch with all significant others. Today, in most organizations use of electronic mail or equivalent tools for work-related communication is not widespread.

If the basic constraints on where and when people work are removed, two questions are important: What is possible? What will happen? It is important to note that most discussions of telecommuting really focus on what is possible rather than current reality.

4. RESEARCH ON CURRENT TELEWORK TRENDS IN THE UNITED STATES

The current body of research on actual telework trends in the U.S. is relatively small, particularly when compared to the plethora of speculative or anecdotal articles about the subject. In this section, relevant research in the U.S. which deals with actual trends rather than speculation is discussed.

4.1. Research on Industrial Homework

Work at home is not new. There is a tradition of labor-management struggle in the U.S. of which work at home is clearly a part. In the 1930's a major impetus of the Fair Labor Standards Act was protection of women and children against exploitative labor in the home. An insightful study by Boris [2] compares the struggle over bans on work in the home in the 1930's with the current "right to work" debate surrounding women knitters in Vermont. She shows how the current debate is being connected to "women's rights" but is rather part of a larger reorganization (i.e., deregulation) of the American political economy that would, in her view, "more firmly entrench the sexual division of labor".

The general debate over industrial home work focuses on the conditions and employment status of the home worker: it is usually assumed that the work involves relatively unskilled tasks paid on a piece-rate basis. Often the worker is on independent contract status rather than employee status, which has implications for benefits and basic protections. Since bans have been placed on home work in certain specific industries, some home work is performed illegally. A recent study of the apparently large "subterranean" industry of home sewing in Canada [13] documents widespread abuses of home workers by their employers. These include low pay, lack of benefits, imposition of unrealistic deadlines, and lack of enforcement of government regulations. Industrial home work is viewed negatively by labor unions, as indicated by the following:

> All clothing workers are threatened by the existence of one category of workers who work under substantially poorer conditions than do the regular labour force. The rise in the prevalence of homework is a symptom of the weakening power of the labour unions in this country [13, p. 6].

4.2. Employment Status

Much of the debate surrounding industrial home work focuses primarily on the employment status of the worker. There is a prevalent assumption that a person working at home is equivalent to an independent contractor; companies that have set up home work programs on that assumption are targets for accusations of worker exploitation. Two companies in particular have received recent attention. Cal Western is being sued by eight former home workers for unfair labor practices (they were defined as independent contractors). Wisconsin Physicians Service was cited in recent testimony in the U.S. Congress [4]. In both cases the independent contractor status was combined with lower pay than on-site employees doing the same work and few or no benefits. In the second case, the move to home work was further seen as a way to circumvent the union, since the company was going through painful labor disputes. According to Costello:

> The WPS case exemplifies the potential for abuse in home-based clerical work. Without regulations preventing companies from replacing more expensive (and unionized) in-house personnel with cheaper, non-union homeworkers, both groups of women stand to lose [4, p. 10].

Moving workers from employment to contractor status certainly has advantages for employers, primarily in giving them the ability to easily expand and contract the labor force with supply and demand. There is evidence that contract work is a growing trend in offices as well as production work [20]. Although there are costs associated with turning to this "external market", the tremendous recent growth in temporary employment agencies should force these costs lower. In 1984 alone payrolls for temporary agencies were $6 billion, an increase of 33 percent in a single year [27, as cited in 20].

But the issue of independent contractor status cannot be dismissed as purely an exploitative tactic on the part of management to reduce labor costs. In an extensive study of home-based clerical workers, Christensen [3] found that:

> Organizational status overrides occupation when work is done at home. For example, self-employed word processors exercise more autonomy than do employed programmers. Word processors, who would be treated as clerical workers in an office, identify themselves as professionals when they own their own word processing companies at home.

4.3. Telecommunications / Transportation Tradeoffs

A different stream of research took place in the mid-1970's, partly in response to the then-pressing energy crisis. The basic premise of this work was the following: Since telecommunications can substitute for transportation of the worker (thus "telework"), significant savings in energy costs can be ensured if steps are taken to facilitate this substitution. The best examples of this work are [21] which developed alternative scenarios and their implications for a single large firm, and [10], which elaborated on different scenarios and showed their potential effects on nation-wide energy savings. The latter report ended with suggestions for public policy initiatives to bring about such substitutions. A retrospective view of this work and its influence is found in [15].

One problem with this work has been the asumption of technological determinism. The Harkness study states that if fifty percent of all office workers worked in or near their homes six out of every seven working days, the savings in fuel consumption from reduced commuting would be about 240,000 barrels of gasoline daily in 1985 [10, p. 111]. The statement refers to technological potential, but the research has been criticized (perhaps unfairly) because these changes did not come about. The problem of assumed technological determinism pervades many popular stories of telecommuting, and so is an important issue to address. A more appropriate view is one of contingencies:

> The conclusion from comparing many studies is that information technologies can indeed encourage and also substitute for the physical movement of goods and people, with consequences for centralization and decentralization. Which of the two effects will appear in any given case appears to depend more on factors other than the choice of technology [17].

4.4. Census Figures

One issue of continuing uncertainty is the lack of accurate figures on how many people work at home as "telecommuters" as well as how many work at home under any conditions. One frequently cited estimate holds that there are currently 10000 teleworkers in the United States [8]. This estimate may appear low until one considers that the implied definition of teleworker is the restricted one used in this paper (employees working at home with information technology on a regular basis as a substitute for commuting to the office). The basis for the estimate is not given in the ESU report.

Census Bureau figures do report on whether the home is the primary place of employment; a steady decline in this number primarily represents a significant decline in the number of farm workers. As reported in [16], only 3.5 percent of workers over 16 worked at home in 1970, and this figure declined to 2.3 percent in 1980. Of these, only 32 percent lived in urban or suburban (non-farm) locations. Pratt [28] reports that in 1980 there were one and a half million home-based businesses and three-quarters of a million people working at home as employees; she does not indicate what constitutes an "employee" under this interpretation of census data.

Thus census figures give no evidence of a significant growth of home workers using information technology.

4.5. Research on Office Location

A related issue to the actual numbers of people working at home is the pattern of shifts in location of offices in relation to the potential labor pool. In a study of back offices in the San Francisco Bay Area, Nelson [20] shows that companies locate back offices in a relatively narrow geographical band where the demographics of the population are highly constrained. The important trend she documents is that the new back office jobs she studied require a specialized and relatively rare set of qualities, and offices are constrained to locations where those qualities can be found. This argument is opposite to the normal thinking that back office decentralization goes hand-in-hand with deskilling and the search for a lower-skilled, cheaper labor pool.

Dahmann [5] reports data on population migration showing that after years of movement from cities to rural areas, from March 1983 to 1984 there was a shift of 351,000 jobs in the opposite direction. He also shows that average commute times increased only very slightly from 1975 to 1980. Dahmann concludes that although people continue to relocate, the jobs are relocating as rapidly as the people are.

These studies show that, like the evidence on employment status, the issue of the motivation of an employer to relocate jobs is a complex one. There is no systematic (at least no successful) effort under way to deskill, reduce pay, and reduce dependence on the clerical work population through judicious use of information technology. Work at home would be a natural extension of this effort; it too is significantly more complex.

4.6. Work / Nonwork Conflicts

Much of the research focusing specifically on experiences with work in the home has been concerned with the relationship between the work and nonwork domains. In particular, the question of whether work at home is an acceptable method for combining income-producing activity and child care is examined.

The most significant work to date related to this domain is by Christensen [3]. In her survey of 215 women working at home with computers, supplemented by detailed interviews of 24 home workers, Christensen concluded that:

> Women who work at home as a way of balancing child care and paid employment typically lie in traditional two-parent households, where the father is the major breadwinner. These women work part-time, primarily for "bonus" money and the psychological benefits of doing something other than being a full-time home-maker and mother. On average, they contribute well below 25 percent of the household income.

Christensen also concludes:

> Women do not work and care for their children simultaneaously. They most often work when their partners can care for the children, or when their children are at school or asleep. When a professional woman has dependable, steady work, she is apt to employ paid child care, in the home or outside.

In my own survey [25] in which professional and clerical women were interviewed, I found that using work at home as a means of simultaneously working and providing child care has certain negative aspects which should not be overlooked. These women experienced a frantic pace of activity with constant stress and pressure from both work and family demands and little time for themselves or for leisure activities. Not surprisingly, the women with children consistently reported increased stress associated with work at home, regardless of whether they had supplemental child care. These women felt they were constantly juggling a complex schedule of activities, and were being pulled by the simultaneous and conflicting demands of work and family roles. This exploratory analysis raises the question of whether work at home is beneficial for the employee with primary child care, or dysfunctional to both work performance and child care as well as highly stressful.

The debate over the inadequacy of current child care alternatives and whether work at home is an acceptable alternative continues to be an active one. As Boris [2] points out, the current administrative efforts to deregulate work at home are based on an appeal to a combination of traditional values of child care with the right to work: Women have a right to work and take care of their children at the same time. What little research there is to date seriously questions whether this combination is a right or a burden borne primarily by women.

4.7. Computer Use in Households

There has been a small amount of research on use of personal computers in the home that bears some relationship to the topic of this conference. In a survey of 282 home computer users, Vitalari et al [31] found that approximately 45 percent of computer use in the home is spent on work-related activities. They conclude that "home computers engender a shift from recreational or pleasure-oriented activities (e.g., television watching) to task-oriented activities .... The household of the future may be the site of more task-oriented behaviors." Their sample was drawn from computer clubs and was heavily oriented toward early adopters and those in technical professions; it may have been that these people had more justification for a personal computer based on task-oriented needs, and were thus motivated to purchase one sooner than the rest of the population.

In a similar vein, Horowitz [12] concludes from her research on computer use in the home that there is a preliminary trend to seeing the household as an income-producing unit.

4.8. Research on Telecommuting

Although several of the studies discussed above concerned work at home with computer technology, their primary focus was not on telecommuting per se. In this section, studies whose primary concern was the relationship between the employee and the employer and the feasibility of telecommuting as a permanent employee work arrangement are reviewed.

McClintock [18] interviewed twenty telecommuters to determine the effects of their work arrangement on their productivity. He found they experienced greater productivity on routine tasks, primarily because of access to an electronic mail system. His respondents also felt they increased their effectiveness on complex tasks because of fewer interruptions. They felt, somewhat surprisingly, that they had greater interdependence with coworkers and more effective use of face-to-face contact as a result of their home work arrangement.

In my own exploratory survey [22], I interviewed ten employees who were geographically separated from their managers at least part of the week; I also interviewed their managers. All the employees were professionals working on long-term deliverables. I found a tendency to formalization of supervision of the remote employee, possibly representing differential treatment. Managers acknowledged that remote supervision was more time-consuming; they also depended on selection of employees who were already highly motivated and self-disciplined, that the manager could trust to be productive. Even so, managers admitted to being uncomfortable not being able to "see" their employees working.

Other studies report on particular companies or experiments. Kraut [16] conducted a survey of professionals at Bell Communications Research, and concluded that "Overall, the time people spend working at home is independent of the time they spend working in the office." He concludes that telecommuting is not a significant phenomenon, the primary reason being "incompatibility with the current work ethos."

In longitudinal evaluations of three corporate telecommuting experiments, I found noticeable differences between employees and managers in their perceptions of the effect of telecommuting on work performance. In general, employees felt that the opportunity to work at home several days a week enhanced their work performance, improved their personal job satisfaction, and had no negative effect on their chance for promotion or their relationship with their supervisors. Managers, on the other hand, were considerably more conservative about the effect of the arrangement on employee productivity. They generally felt that the arrangement was not detrimental to performance but that it entailed significantly more work for them in terms of preparing work assignments and monitoring progress. In general, they considered the work arrangement necessary or tolerable but in virtually all cases would have preferred to have the employee on-site full time if possible.

In a recent survey of 210 life insurance companies, Moore [19] shows that only a handful are currently involved in telecommuting. Most reported incidents are in addition to regular work hours; most are informal and random, and companies have no formal policy regarding telecommuting as an employee work option. A recent survey of Canadian companies [14] shows that, although they recognize the need to provide employees with flexible work scheduling alternatives, only 4.5 percent of those responding had any kind of home work program. In response to a request for more information, only 15 percent wanted to know more about home work, the lowest response of six categories of employee work options.

4.9. Conclusions from the Research to Date

Based on the research reviewed in this section, the question of whether telecommuting is a significant phenomenon today certainly cannot be answered clearly in the affirmative. One point is certain: information technology is not the driving force [3,16,17,20,22,23,24,25]. Information technology may make new forms of work organization possible, but organizational culture as well as economic and social concerns of employees and employers have a stronger influence over what choices are actually made.

A second point is also clear: Telecommuting is not necessarily favored by management. In fact, it is quite the opposite: most managers, given the choice, prefer to "see" their employees, and for them telecommuting is more of a hassle than a benefit.

5. A DEMOGRAPHIC SURVEY OF TELEWORKERS

In this section, the results of two magazine surveys on telework are discussed. The purpose of the surveys was to document the extent of the trend to telework in a population which is presumably doing so today. For this reason, a random sample of U.S. households or U.S. workers (even office workers) was not feasible. It was decided instead to target magazines whose readership best fits those who appear to be telecommuters under the best of circumstances.

Two trade magazines were chosen for the demographic survey, one targetted to computer professionals and one on personal computing for general professionals. The survey was sent by mail to a random sample of 5000 subscribers to each magazine. The results from the combined sample are summarized below.

Table 5.1 shows the number of respondents who reported working at home. (The survey also asked for the opinions of persons not currently working at home.) The remaining tables report only on that portion of the sample.

5.1. Results

TABLE 5.1
Do you work at home?

	FREQ	PCT
Yes	807	50.1
No	805	49.9

TABLE 5.2
When you are working at home, what is your employment status?

	FREQ	PCT
Employed by a company or another person	342	42.4
Self-employed	351	43.5
Other	114	14.1

TABLE 5.3
How are you paid for the work that you do at home?

	FREQ	PCT
Salary	310	38.4
Commission, contract, etc.	134	16.6
Profits	110	13.6
Hourly or daily	105	13.1
Piece-rate	38	4.7
Other	110	13.6

TABLE 5.4
How much of your income is provided by your work at home?

	FREQ	PCT
Less than 25%	525	65.1
25 - 49%	77	9.5
50 - 74%	37	4.6
75 - 99%	21	2.6
100%	63	7.8
Other/no response	84	10.4

TABLE 5.5
Hours Worked

How many hours do you work in an average week?	50.6 hours avg.
How many of those hours do you work at home?	14.7 hours avg.

TABLE 5.6
Are the hours that you work at home:

	FREQ	PCT
In addition to regular work hours	469	58.1
An occasional substitute for work at another location	97	12.0
A regular substitute for work at another location	95	11.8
All the paid work you do	87	10.8
Other/no response	59	7.3

TABLE 5.7
What would be your ideal work arrangement?

	FREQ	PCT
To work part-time in my home, part-time outside	535	66.3
To work only in my home	126	15.6
To work entirely outside of my home	53	6.6
Other	93	11.5

TABLE 5.8
Overall, how satisfied are you working at home?

	FREQ	PCT
Very satisfied	407	50.4
Somewhat satisfied	282	35.0
Somewhat dissatisfied	43	5.3
Very dissatisfied	4	0.5
No response	71	8.8

TABLE 5.9
Why did you first decide to work at home?
(Respondents gave multiple answers.)

	FREQ	PCT
To increase my productivity	414	51.3
To work in my own way, at my own pace	390	48.3
To earn extra money	266	33.0
To save time commuting	160	19.8
Tax benefits	129	16.0
Low overhead	124	15.4
Other	124	15.4
To ease conflicts between work and family	110	13.6
To take care of family	63	7.8
To avoid office politics	61	7.6

TABLE 5.10
What are the advantages of working at home?
(Respondents gave multiple answers.)

ADVANTAGE	FREQ	PCT
More productivity	499	61.8
More time with my family	290	35.9
More time to myself	263	32.6
More money	223	27.6
Increased career opportunities	185	22.9
Less personal conflict	116	14.4
No advantages	18	2.2

TABLE 5.11
What are the disadvantages of working at home?
(Respondents gave multiple answers.)

DISADVANTAGE	FREQ	PCT
Lack of interaction with co-workers	269	33.3
Work too much	258	32.0
Less time to myself	134	16.6
Less time with my family	82	10.2
Resentment of my spouse	69	8.6
Increased stress	63	7.8
No disadvantages	151	18.7

TABLE 5.12
Which type of computer equipment do you have at home for your work-related use?

	FREQ	PCT
Personal computer	588	72.9
Modem	309	38.3
Word Processor	151	18.7
Terminal	150	18.6
Other	54	6.7

TABLE 5.13
Who owns the equipment?

	FREQ	PCT
My family or I do	493	61.1
Employer or client	119	14.7
Other	195	24.2

TABLE 5.14
Demographics

SEX	FREQ	PCT
Male	678	84.0
Female	127	15.7
No response	2	0.2

MARITAL STATUS	FREQ	PCT
Married. or living with partner	666	82.5
Divorced, or widowed, or single	141	17.5

HOUSEHOLD INCOME	FREQ	PCT
Under $30,000	98	12.2
$30,000 - 59,999	438	54.2
$60,000 and over	250	31.0
No response	21	2.6

AVERAGE AGE: 42.5 years

AVERAGE NUMBER OF CHILDREN: 2.1

5.2. Discussion

Is telework a significant departure from the daily commute to a nine-to-five workplace? The data shows that the respondents, like others in similar professions, work long hours. Although the average number of hours worked at home is equivalent to nearly two work days, most them appear to be worked in addition to regular work hours. It appears that the one significant change in work habits that has come about is that now an employee can perform (unpaid) overtime work in the convenience of one's home and surrounded by one's family, instead of having to stay long hours at the office in order to access the equipment.

Why did they decide to work at home? Clearly this group is motivated to increase their productivity. Whether they find the office too distracting or are worried about not getting enough work done or are constantly under deadlines, they choose to extend their work day into their home life in order to get more work accomplished. It is fairly clear that for the most part they are not compensated directly by employers (i.e., overtime) for the work they do at home. They are also not motivated by family considerations, although many seem to feel that it is a better choice to be near one's family while working than working longer hours at the office (1). They may feel that this way they can share regular meals with their families and be physically present in the evening hours, even though they might be off in a separate room toiling over their terminals while the rest of the family watches television.

How do they like working at home? Clearly many feel that their goal of increasing their productivity is accomplished. Of course this result must be considered with caution, since strictly speaking productivity is output per unit of input (hours worked) and they may be simply extending their hours rather than increasing their output per hour. However, interviews with people who work at home show they do feel that their actual productivity has increased, due to the lack of distractions and interruptions in the home environment [24].

Working at home is not ideal. The most frequently cited disadvantage is lack of interaction with coworkers. This is a particularly interesting result, given that readers of one of the magazines are computer professionals. The myth that programmers are solitary types, preferring their terminals to people and thus ideally suited to working in the solitude of the home, has basically been dispelled. In fact, they are very social types, with a primary topic of converation being how to use their computers. Thus an important part of learning and professional development of computer professionals is constant interaction with peers, and they miss that when they work at home.

Secondly, this group of people tends to work too much, and least some recognize that the convenience of the equipment in the home brings the disadvantage that they tend to use it, sometimes causing family conflict. This problem has often been stated in interviews [25]. The terminal or computer is close and inviting, and it is tempting (particularly with electronic mail) to just sign on and "check my mail" or "see who else is on the system". The productivity benefits cited above have this downside in that the presence of the machine is compelling.

Overall, however, those who work at home and responded to this survey seem to feel that the advantages outweigh the disadvantages. Over 85 percent reported being at least somewhat satisfied with the opportunity to work at home. They do not want to work at home full time, as is apparent from Table 5.7. The overwhelming majority favor the flexibility to be able to work at home part of the time but still have a regular workplace outside of the home.

This is a homogeneous group. Most are male and married (no data was collected about spouse's occupation); eighty-five percent earn at least $30,000 per year (although the question was stated in terms of household income which also includes spouse's income); thirty-one percent earn over $60,000.

5.3. Conclusions from the Demographic Survey

Does the data indicate that a dramatic shift in work location, from central offices to "electronic cottages", has taken place? The answer is clearly no. Instead, information technology has made it easier to increase the total number of hours worked by allowing work at home to substitute for what might have been longer hours in the office.

Clearly, the respondents to this survey are a privileged group in terms of employment status. The jobs they do at home are those that generally enjoy a significant degree of autonomy and have been performed at least partly in the home without technological support. Those who work at home, even in addition to regular work hours rather than as a substitute, choose to do so because of the autonomy to work at one's own pace and to thus benefit from increase productivity. The large majority have a spouse who lives with them, and although we did not ask if the spouse works outside of the home, it is clear that only a very small percentage of the respondents work at home even in part in order to help with child care.

For those who work at home in this sample, the advantages far outweigh the disadvantages. Since most do not work exclusively in the home, the disadvantage of lack of interaction with coworkers is probably not critical. However, having access to equipment and work-related materials in the home may encourage them to work too much. Indeed, they work long hours and otherwise show signs of being "workaholics".

6. FUTURE DEVELOPMENTS IN TELEWORK IN THE UNITED STATES

Based on the discussion above, it should be clear that I do not hold to the prediction that telework, as defined and practiced today, will become a prevalent form of work organization in the future. My basic conclusion is that while telework may be technologically feasible in the near future, it is not technologically driven. Indeed, organizational culture and traditional bureaucratic structure are strong inhibitors. In the long run, a combination of technological, economic, and social forces may bring about significant changes in organizational structure and culture, which may in turn lead to new work organizations including remote collaboration and remote supervision. In this section, these conclusions are elaborated in terms of technical, economic, organizational, legal, and social / psychological forces.

6.1. Technical Forces

Earlier in this report the current state of information technology with respect to support for telework was discussed. It was pointed out that while telework as a phenomenon is not driven by technology, the current state of technology represents a constraint rather than a facilitator. This is primarily because the use of information technology for interpersonal communication is not widespread in organizations today. However, I predict that significant developments in the use of technology to support and add value to interpersonal communication, particularly in work-related communication and collaboration, will occur in the near future. A recent conference (December 1986) entitled "Computer Support for Cooperative Work" marked the beginning of a trend in Computer Science research labs and universities to development of computer-based tools for work groups. As an example, researchers at Xerox Palo Alto Research Center are experimenting with video and computing to support their own remote collaboration; half of the research laboratory is located 400 miles away in Portland, Oregon [9].

It is my belief that when such tools become cost effective, and as the costs of telecommunications to support wide bandwidth communications decrease, remote collaboration will become commonplace. By this I mean that it will become relatively easy for people (i.e., professionals) to work together even when they are not co-located. Organizations, particularly those which are already physically decentralized, will quickly take advantage of the opportunity to utilize scarce specialized human resources effectively without incurring travel expenses.

Clearly, remote collaboration also requires remote supervision. To date, issues involving the process of remote supervision (e.g., training) and technological support for it (e.g., work measurement, communication support) have not been addressed by either computer science or management researchers.

When the technology to support remote collaboration and remote supervision are in place, will it matter where people work? Will most of them be at home? It is my opinion that economic and social forces work otherwise (as discussed below), but certainly telework will be more feasible than it is with today's technology.

6.2. Economic Forces

What major economic forces will work to encourage or discourage telework in the U.S.?

For employees, the two-income family will continue to be a fact of life. In order to support nonwork requirements (i.e., family) as well as two jobs or single-parent families, employees will continue to look for flexibility as an important criterion in their work, even sacrificing income to attain it. With day care in the U.S. completely inadequate and prohibitively expensive, the cost of going to work, when clothes, commuting, and incidentals are also factored in, is considerable for a second wage-earner. Will employee demands for flexibility in work schedules be heeded? Only if skilled employees are in sufficiently short supply that organizations have no alternative. Unfortunately for employees, for all but a few white-collar occupations (such as software specialists), this is not the case.

On the other hand, the economics of environmental uncertainty and competitive pressure will continue to force organizations to look for ways to respond quickly to changes in demand for products and services. This means finding ways to quickly expand and contract output, and an important factor will be the ability to quickly expand and contract the labor force. This can be accomplished most efficiently through an increase in contract work, particularly as transaction costs for acquiring contracts, distributing work, and measuring output are reduced due to information technology [32].

6.3. Organizational Forces

Given the environmental forces described above, I fully expect organizations to increase the substitution of contract labor for full-time employment in more and more domains of office work, professional as well as clerical. This trend alone may signify an increase in telework, since the employer does not bear the cost of space to perform the work. As noted below, I believe a significant issue requiring immediate attention is protection of contract labor.

Organizations will also continue a trend to physical decentralization, facilitated but not driven by lowered costs of telecommunications. Facilities location is now dictated by costs of real estate and energy as well as location of an adequate work force. In the U.S., continued "suburbanization" of back-office facilities as well as "off-shore" work should be expected. In many cases today, location of a facility in a suburban location with an adequate population of skilled employees (primarily full-time homemakers with few or no alternative work options) solves the problems of shortages of skilled employees at least temporarily and renders telework as an employee work option unnecessary.

6.4. Legal Forces

How will the legal status of telework change in the near future? Today, there is polarization with respect to attitudes toward telework. The current administration supports complete deregulation of all forms of work at home; other branches of congress, supported by the unions, are working actively to extend current bans on home work to work performed on a computer. There are significant problems with respect to enforcement of such a ban.

As shown in the recent Congressional hearings on home work, a recurring important theme is the status of the worker who is working at home. If the worker is considered a full-time employee, he or she has the right to receive full benefits and pension rights. If the worker is an independent contractor, no such rights exist. In current laws, there is a large "gray area" concerning part-time employment and piece rates; organizations such as Blue Cross-Blue Shield are testing the gray areas of the employee protection laws with their home work programs. It is my strong belief that work at home is not the issue here; rather, the issue is clarification of the rights and privileges of employee versus contract status in employee compensation. The definition of an independent contractor is an important issue to be addressed through legal and legislative channels, over and above the issues related to protection of home workers specifically. I believe that the next Administration, following the presidential elections of 1988, will be in a position to address this issue.

6.5. Social / Psychological Forces

It was noted that employees will increase their demands for flexibility in their work lives, in order to accommodate nonwork demands. It is my belief that the problem of attempting to combine full-time employment and full-time child care, whether or not the employment takes place in the home, is paramount. Many writers have speculated on the social isolation of people working at home; evidence shows that this problem is much less important than the significant strains produced from trying to work and take care of children simultaneaously.

However, the one overlooked social force which is and will continue to be a major barrier to telework is organizational culture, and in particular management style. Managers do not like telework; they want to see their employees. Over and over again I have heard the complaint "How can I manage someone I can't see?" Having an employee work effectively at home requires management skills such as trust and confidence in the employee's abilities -- skills which represent good management in any case. A bad manager does not want his or her inadequacies exposed, and having an employee working at home increases the manager's vulnerability to exposure of poor general management practices.

Furthermore, organizational culture dictates a commitment to the organization as a place. Companies incur tremendous expense providing facilities in which an employee can feel a sense of belonging and safety, with health facilities, libraries, natural surroundings, as well as cafeterias and parking lots. Such trappings are designed to keep employees "in"; once they walk over the threshold they "belong" to the organization until they leave. Employee productivity, particularly of professional workers, is measured by time in rather than output. Signs of status abound, from the size of one's office to whom one speaks to in the elevator. Promotability is keyed above all else to visibility, not to performance on some objective criteria. While this approach to employee performance and control may appear on the surface to be inefficient, it is entrenched in bureaucracies and changes slowly. In all respects, the notion of an employee working at home when and where he or she wants flies in the face of this corporate culture.

7. CONCLUSIONS

My conclusions address the question, "What strategies can be developed to stimulate useful and desirable organizational forms of telework?"

I don't know that there are any desirable organizational forms of telework today. There are many arguments against telework, primarily in terms of its potential to be used to exploit workers. In these arguments, organizations are usually described as poised and ready to implement telework in exploitative ways as soon as certain legal barriers are removed. I hope this paper has demonstrated that, in the U.S. at least, this is simply not the case. Organizations are NOT particularly interested in telework as an employee work option. Furthermore, the technological forces have not been fully developed, so that from a technical standpoint telework is still difficult for most work.

I believe that as technical developments encourage remote collaboration and remote supervision, telework will take on a different meaning, not focused on work location "in or out" of the organization. Physical organizational boundaries will become less clearly defined in a general way. The definition of "employment" will also become less clear as part-time and contract work become commonplace. These trends will override telework. I believe there will be a slow migration in work toward telework. Two issues are of paramount importance in the United States, regardless of whether they result in an increase in telework. These are improvements in affordable, adequate day care alternatives and protections for work under independent contractor status.

8. FOOTNOTES

1. The author is currently working with Professor Christensen, who developed the original questionnaire, to compare this sample with the sample from Family Circle Magazine, whose readership is primarily women earning second incomes. The differences between the two samples are expected to be dramatic.

9. REFERENCES

[1] Applegath, J. Working Free, Washington DC: World Future Society Press, 1982.
[2] Boris, Eileen, "'Right to Work' as a 'Women's Right': the Debate over the Vermont Knitters, 1980-1985", Series 1, Legal History Progam Working Papers (LHP-#1:5), Institute for Legal Studies, University of Wisconsin - Madison Law School, February 1986.
[3] Christensen, Kathleen, "Impacts of Computer-Mediated Home-Based Work on Women and Their Families", Graduate Center, City University of New York, June 1985.
[4] Costello, Cynthia, "The Office Homework Program at the Wisconsin Physicians Service Insurance Company", Hearing Record, Pros and Cons of Home-Based Clerical Work, House of Representatives, Employment and Housing Subcommittee of Committee on Government Operations, February 26, 1986 (Doc. #58-9460).
[5] Dahmann, D., Geographical Mobility: March 1983 to March 1984, U.S. Department of Commerce, Bureau of the Census, Series P-20, No. 407, September 1986.
[6] Diebold Automated Office Program, Office Work in the Home: Scenarios and Prospects for the 1980's, New York: The Diebold Group, Inc., August, 1981.
[7] Edwards, Paul and Sarah, Working From Home: Everything You Need to Know about Living and Working under the Same Roof, Los Angeles: J.P. Tarcher, 1985.
[8] Electronic Services Unlimited, Telework: A Multi-Client Study, New York, 1984.
[9] Goodman, G. and Abel, M., "Collaboration Research in SCL", Proceedings, First International Conference on Computer Support for Cooperative Work, Austin, TX, December 1986.
[10] Harkness, R.C., Technology Assessment of Telecommunications /Transportation Interactions, Menlo Park, CA: SRI International, May, 1977.
[11] Hewes, J.J., Worksteads: Living and Working in the Same Place, New York: Doubleday, 1981.
[12] Horowitz, Jamie, "Working at Home and Being at Home: The Interaction of Microcomputers and the Social Life of Households", PhD dissertation, Dept. of Environmental Psychology, Graduate Center, City University of New York, 1986.
[13] Johnson, Laura C., The Seam Allowance: Industrial Homework in Canada, Toronto: The Women's Press, 1982.
[14] Johnson, Laura C., "Working Families: Workplace Supports for Families", Social Metropolitan Planning Council, Toronto, Canada, 1986.
[15] Kraemer, K.L., "Telecommunications - Transportation Substitution and Energy Productivity: A Re-Examination," Telecommunications Policy, Vol. 6 (1), March 1982, pp.39-59, and Volume 6 (2), June 1982, pp. 87-99.

[16] Kraut, Robert E., "Telework: Cautious Pessimism", R. Kraut, editor, Technology and the Transformation of White Collar Work, Hillsdale, NJ: Erlbaum, 1986, pp. 135-152.
[17] Mandeville, Thomas, "The Spatial Effects of Information Technology: Some Literature", Futures, February 1983.
[18] McClintock, C. C., "Working Alone Together: Managing Telecommuting," National Telecommunications Conference, December 1981.
[19] Moore, Kay, 1986 Personnel Policies And Practices Survey, Atlanta: Life Office Management Associate, 1986.
[20] Nelson, Kristin, "Automation, Skill, and Back Office Location," paper presented to Association of American Geographers, Minneapolis, MN, May 1986.
[21] Nilles, J.M., Carlson, F.R., Gray, P., and Hanneman, G.G., The Telecommunications-Transportation Tradeoff, New York: John Wiley and Sons, 1976.
[22] Olson, M. H., "New Information Technology and Organization Culture," Management Information Systems Quarterly, Vol. 6 (5), December 1982, pp. 71-92.
[23] Olson, M. H., Final Project Report, NSF Grant No. IST-8208451, 7/182-3/3/83, March 1983.
[24] Olson, M. H., "Remote Office Work: Changing Work Patterns in Space and Time", Communications of the ACM, Vol. 26 (3), March 1983, pp. 182-187
[25] Olson, M.H. and Primps, S.B., "Working at Home with Computers: Work and Nonwork Issues", Journal of Social Issues, Vol. 40 (3), 1984, pp. 97-112.
[26] Olson, M.H., "Do You Telecommute?", Datamation, October 1985.
[27] Pfeffer, J. and Baron, J., "Taking the Workers Back Out?: Recent Trends in Labor Contracting", paper presented for Stanford Graduate School of Business / Business and Social Research Institute Joint Workshop, Stockholm, Sweden, September 1985.
[28] Pratt, J.H. and Davis, J.A., "Measurement and Evaluation of Family-Owned and Home-Based Businesses", U.S. Department of Commerce, National Technical Information Service, Springfield, VA, July 1986.
[29] Sproull, L., Kiesler, S., and Zubrow, D., "Enountering an Alien Culture", Journal of Social Issues, Vol. 40 (3), 1984, pp. 31-48.
[30] U.S. Government Printing Office, "Pros and Cons of Home-Based Clerical Work," Hearing before a Subcommittee of the Committee on Government Operations, House of Representatives, February 26, 1986.
[31] Vitalari, N.P., Venkatesh, A., and Gronhaug, K., "Computing in the Home: Shifts in the Time Allocation Patterns of Households", Communications of the ACM, Vol. 28 (5), May 1985, pp. 512-522.
[32] Williamson, O.E., Markets and Hierarchies: Analysis and Antitrust Implications, New York: Free Press, 1975.

Social aspects of telework:

FACTS, HOPES, FEARS, IDEAS

Marilyn Mehlmann

In 1982 Bo Hedberg and the author wrote a paper (1,2) on the subject of telecommuting. It was primarily speculation, a piece of fantasy in which we contrasted working at home with working at a neighbourhood work centre, or corner office, largely to the detriment of the former.

Since then a lot has happened - qualitatively but not, it would seem quantitatively. That is, experiments and experience have answered some of the questions we posed then. But the number of people actually engaged in telecommuting is still so small as to be practically invisible. Or at least this seems to be the case but the lack of reliable statistics is striking, and probably indicative of the difficulties in fitting this phenomenon into our current frame of reference. But it should be possible by now to examine our original speculations, accept or discard them, and perhaps partly replace them by new hopes, fears and ideas.

WHAT WE THOUGHT IN 1981

In summary, in our paper "Computer Power to the People" (1,2) we postulated that telecommuting posed a threat of polarization: the "haves" would experience it as yet another fringe benefit whereas the "have-nots" would be liable to lose a part of what little they had (all in accordance with the New Testament (3)).

	Sink group	Float group
Professional skills	Low	High
Status	Low	High
Bargaining power	Low	High
Work involvement	Part-time	Full-time
Terminal-based work	Full-time	Part-time
Voluntary move	No	Yes
Payment	Low paid	Well paid
Residential area	Few contacts with neighbours	Many contacts with neighbours
Professional network	Poor	Rich
Unionization	Low	High
Social environment	Poor	Varied

Table 1. Working at home - two extremes

We did however think that collective work places in corner offices could, while not guaranteeing a positive outcome, perhaps alleviate many of the potential problems of home working and at the same time make a positive contribution to the local community.

Dimension	Computer resource centre	Home terminals
Social contacts	Many, face-to-face	Few, man-computer
Learning environment	Rich, varied, many cues/examples	Poor, limited, few cues/examples
Cross-cultural	Yes	No
Per capita investments	Moderate	High
Utilization	High	Low
Self-determination over workplace	Limited	High
Work organization	Many combinations	Very few alternatives
Core unit	Local community	Nuclear family
Compatible with	Institutional solutions (day-care centre, grocery, local bank, newspaper, library)	Private solutions (home shopping, electronic news, home banking, family day-care)
Integration between work sphere and private sphere	Separate but close	Totally integrated or totally separated
Labour divisioning, trends specialization	Less pronounced than today	Continuing present
Compatible with union structure/operations	Yes	No

Table 2. Comparison between the two scenarios in some selected dimensions

THE CORNER OFFICE IN NYKVARN

The Nykvarn project was a research project funded by the Swedish Council for Building Research and the National Bank of Sweden Research Fund. It was carried out under the auspices of a Nordic social research institute. It attracted an enormous amount of interest in the Nordic countries, not only from research workers but also from the public, the press, television, and to some extent from employers. It has been documented in at least one report in English (4) and another in Swedish (5).

It is perhaps a measure of the general lack of serious experiment and experience that this rather modest project, involving a handful of telecommuters, attracted so much interest and received so much support.

From the point of view of the individuals concerned, the project team, and the local community, Nykvarn can according to the reports be regarded as an almost unqualified success. Those problems reported are represented as minor, and in no way comparable to the considerable benefits achieved.

The Nykvarn corner office displayed several of the dynamic properties we postu-lated in our paper (1), despite an initial handicap: the project was not a local initiative as we presupposed but a proposal from a number of research workers and organizations. Its origins were therefore artificial, and indeed a number of the people working there actually commuted *to* Nykvarn, which in some ways negated the purpose. The reports do not comment on whether these factors could in any way contribute to explaining the low proportion of working time spent at the centre by those participating: the median value was around one third of working time (for both full-time and part-time workers), or somewhat less than half of the predicted proportion.

Nonetheless by the end of the research period the composition of this very small work force displayed an interesting mix of telecommuters with their main place of work elsewhere, local employees and self-employed people. All were agreed that the experience had been overwhelmingly positive.

An attempt to relate Nykvarn experiences to our predictions in table 2 yields the following points:

Polarization we did not expect to be a feature of the corner office, nor are any such problems reported. Indeed, primarily "haves" seem to have been selected by their employers to participate in the experiment, apparently because "have-nots" were not believed to be capable of responsibly handling so much freedom. Some employers who were approached declined to participate at all on grounds that seem to indicate a belief that the degree of freedom would be too great for *any* of their employees.

Several of those people who did participate thought that their jobs had been further enriched, or could be if the experiment were to continue. How, if at all, this would affect the work situation of colleauges at the main place of work is not discussed in the reports.

Social contacts: many face-to-face . Somewhat disappointing to participants, who all agreed that the centre should have more work places - at least ten people should spend a large proportion of their working time there, said participants. It is an advantage, they say, when two or more people with the same employer can be there together.

Environment: cross-cultural, rich, varied learning opportunitites . Yes, participants at Nykvarn cooperated and offered each other "work-related reciprocal services" and other forms of cooperation, including joint social arrangements (Christmas lunch, for example). "The absence of a professional hierarchy" is noted as facilitating cooperation and social contacts. Several participants mention that they have gained greater technical competence and greater insight into the potential of IT in their work.

Integration between work and private life ("separate but close"). Few indications, but there is a mention that some participants began to work more flexible hours in order to take advantage of proximity to the home. One father also said that he now accompanied his small daughter to her day care centre and fetched her after work.

Work specialization less pronounced than formerly. Yes, participants "have either retained their earlier tasks or received added tasks... mainly of a qualified nature". Some, however, could carry out only part of their work at the centre, their other tasks being such that they were required to travel to the main office.

Compatible with trade union activity. Yes, at Nykvarn there has if anything been some increase in participants' interest in union questions, according to the reports.

The financial dimensions in table 2 (per capita investment and utilization) are touched upon below. As regards the remaining dimensions no conclusions can be drawn from the reports.

Not only participants but also their employers were positive. Almost all the people working at the centre who were in fact telecommuters (i e employed by a company at some distance from Nykvarn) also had places of work at the main office. Opinions are divided as to whether this is good or bad from the point of view of the employee and the centre. However from the point of view of the employer it clearly means that all costs at Nykvarn (premises, equipment, communications) would have been additional costs. They were in fact borne entirely by the research project.

Now that the research project is over, the premises are charged for (though at lower than normal commercial rates, it would appear); equipment is still provided free by sponsors. The question of financing is clearly not simple. Indeed, in the research reports the conclusion is drawn that the corner office is "not a form of work-place that springs up spontaneously where there is a need". To this subject we return below.

WHAT'S GOING ON?

If Nykvarn is a small, neat, well-delineated, well-researched subject, the opposite is true of almost everything else connected with telecommuting. For example, a recent study by a Nordic social research institute (6) concluded that it is not possible to determine how many employed or self-employed people work in their homes in Sweden. This despite the fact that the population of Sweden is unusually well described, statistically speaking. The report compares in tabular form the meagre results obtained from two surveys, both by the National Bureau of Statistics.

	SURVEY A	SURVEY B
No of people working at home	126 000	340 000
Of whom employed	74 000	140 000

Table 3. Different figures on home working in Sweden

The total number of people in survey B (but not survey A) includes self-employed farmers, who from other sources can be estimated at very roughly 150 000. Nonetheless a large discrepancy remains, as well as a number of questions as to what was actually being measured. Assuming however that the figures do say anything about the prevalence of home working, something very roughly of the order of 100 000 people might be employed but working almost exclusively at home. This out of a population of around 8 million. Some of these will presumably be doing work far removed from IT, for example outdoor manufacturing or telephone sales. Some could perhaps be working *out of* the home rather than in it, e g sales personnel, but this is unclear. Certainly not all can be classified as telecommuters.

The report adds that it was not possible to shed any further light on the matter with the help of the trades union, some of which have views or even policies concerning home working but none of which had any statistics; nor with the help of such State bodies as that for the protection of the working environment, whose jurisdiction does not include home workers. The home is generally excluded from legislation concerning safety and hygiene at work.

Certainly a significant change has already taken place in Swedish society, more radical than perhaps in most other countries. Whereas many people formerly spent their days in or near the home, this has of recent years become a rarity. Infants *go to* day care centres, children *go to* school, adults (both men and women) *go to* work. Generally speaking the distance travelled tends to increase. Only pensioners are left in the residential areas during the day - pensioners, the handicapped, and the unemployed (still a small minority in Sweden). But pensioners and the handicapped often have access to day centres too...

NEW QUESTIONS

It has been widely assumed both that the prevalence of home working was declining to insignificance, and that this is a positive development. Neither assumption is incontestable. But without access to facts both as to prevalence and as to consequences, it is hardly possible to meaningfully intervene in what seems to be a new trend, a trend towards the breaking up of rigid conventions governing the time and place of work (Toffler, 7). It is not even possible to know, for certain, that the trend actually exists or is gaining momentum, though this is frequently claimed at least as regards the United States (Naisbitt, 8).

One thing is certain: since the beginning of work, the time and place for its execution have largely been determined by the exigencies of the work itself. This is as true of the hunter and nomad as of the industrial worker commuting to a factory located near sources of energy or raw materials.

But that is beginning to change. Work is beginning to be relocated to mesh with a very particular source of energy and raw material, namely the work force. This is certainly not (always) done out of regard for the individual worker. The relocation of electronics manufacturing to south-east Asia has more to do with the convenience of shareholders than with that of employees. Other examples are perhaps more interesting as a basis for predicting the spread of teleworking.

An attempt to classify office work according to a number of variables is shown in Figure 1. There is no doubt that the overwhelming majority of white collar workers today belong to Type A, a conventional office environment, even if (as mentioned above) it is extremely difficult to obtain reliable statistics. A survey including the variables shown in Figure 1, and with the added dimension of questions concerning the use of IT, would shed some light on the subject. In our original paper (1), the "haves" were generally of class B and the "have-nots" of class C.

The assumption behind studies and predictions of a spread of teleworking is that some or many type A jobs can be converted to type B or C, though much development to date seems to have been from A to F (11).

Various claims have been made concerning the numbers of Type A jobs that could suitably be carried out wholly or partly away from the regular office. One figure occasionally quoted is 20% (6). An employers' association in Sweden proposed a study based on a "could - would - should" approach:

1. Feasibility. What tasks *could* reasonably be carried out at a distance?
2. Profitability (in the broadest sense): what tasks *would* employers wish to spin off in this way?
3. Desirability: what tasks *should* (from the point of "society") be spun off - if any?

	A	B	C	D	E	F	G
Employed	yes	yes	yes	yes	yes	no	no
Place of work "at work"[1]	yes	yes	no	yes	no	no	(no)
Place of work near home[2]	no	yes and/or	yes or	no	yes	yes and/or	yes and/or
Work in home	no	yes	yes	yes	no	yes	yes
Presence controlled	yes	no	no	no	yes	no	no
Output controlled	no	yes	yes	yes	no	yes	no

A = Clerks, other office workers, most managers
B = Certain specialists, especially those in short supply (e g programmers)
C = Outdoor workers in office production (e g typists)
D = Sales and demonstration personnel, some educators
E = Branch office
F = Self-empoyed suppliers of services (e g document production)
G = Self-employed farmers, consultants, craftsmen and women, artists etc

[1] Locality determined by needs/claims of work
[2] Within comfortable cycling distance (or equivalent)

Figure 1. A way of classifying office work

So far only step one has been carried out. Very briefly, the report (9) identifies 133 different tasks (ordered in six categories) that could be spun off. The list has been developed together with five participating companies, and is surprisingly broad. The report does not give any estimate of how much work might survive the three-step elimination process.

The confederation of white-collar unions, TCO, has since started a project to carry out what might be considered as step 2, but from a union point of view: what kinds of tasks could suitably be carried out at a distance, from the point of view of the employee? Results are expected in 1987 and 1988. The study has begun with an attempt to ascertain the prevalence of teleworking today. By the end of October 1986 replies to the questionnaire had been received from union groups at 69 different work places, five different unions. This preliminary scan has revealed about 170 people working in the home and about 200 in corner offices or equivalent.

As regards home working, so far the typical respondent is a permanently employed man in his 30s who himself took the initiative to start working at home. He often has computer-related work (programming, operations control) and spends more than half his working time at the main office. A typical type B, in other words.

Those in corner offices are also in their 30s. They are marginally more likely to be women than men, and are fairly evenly divided between type B and type C. A bare majority spends her entire working time locally.

FORCES FOR AND AGAINST

Yet another Swedish project, sponsored by the Swedish Centre for Working Life, has made an international study and has described the mechanisms at work motivating various stakeholders to promote or discourage remote working (10, 11, 12).

Clearly the question of costs and financing is interesting. At present we have in Sweden a situation where the tax regulations encourage commuting and discourage home working (6). The costs of commuting are largely tax deductible on a clearly regulated basis whereas any attempt to deduct expenses connected with a working area in the home are regarded with suspicion, treated on a largely ad hoc basis, and frequently refused.

Apart from the role of society, we have a cultural assumption, widely prevalent today in Europe, that the employer should bear the cost of premises, equipment and materials. The employee is normally expected to pay all costs involved in travel between home and work. These assumptions may perhaps need re-examining.

For example, can an employer be expected to pay rent for space used for home working? Can an employee be taxed for private use of a computer placed at home by the company? Or will employees be encouraged to buy or lease their own equipment? And will employers then pay for its use? These questions pertain to home working.
With corner offices, a different set of questions arises. Will employers for instance be expected to provide double work places (space, equipment) for all people working in corner offices? If so the expense is likely to be prohibitive, as noted above. Or can corner office workers share work places - in the corner office, in the main office, or both? Is travel between the two offices to be regarded as "travel in the course of duty" and therefore paid for by the employer, like the trips of sales personnel? Or is it analagous to going to work, and therefore the responsibility of the employee?

If remote working involves extra expenses then it must compete with other costs for the employer. On what grounds could it be justified? The studies sponsored by the Swedish Centre for Working Life identified four major reasons why employers wish to encourage remote working (10):

1. Highly variable work load - home workers are used to cope with seasonal rushes. This is in the report called a *"buffering strategy"*.
2. A *"shedding strategy"* is sometimes used as a first step to dispose of unwanted business activities, groups or individuals.
3. *Recruitment and retainment* of scarce personnel can be facilitated by offering remote working as a fringe benefit.
4. *Better profitability* can sometimes (in some countries) be obtained by exten - sive use of a self-employed work force, reducing the employer's expenses.

In the light of the above, one must support the conclusion of the authors of reports 4 and 5 that positive uses of remote working are not likely to gain much ground if market forces are relied upon.

In reference 12 an attempt is made to set out the pros and cons of remote working from the points of view of the different stakeholders. The results are summarized in Figure 2, parts a-d.

THE INDIVIDUAL (EMPLOYEE)

Advantages
Less commuting, lower costs for travel to and from work
    More time for home and family
More freedom
Flexible working hours
Chance to stay in an area with unemployment
Chance to retain job despite moving to another area
More flexibility

Disadvantages
Isolation (type C)
Chances of personal development or promotion reduced (type C)
Need for personal discipline
Work spills over into free time
Less suitable working environment, unergonomic equipment/situation
Piece rates (type C)
Less job security, fewer benefits
More specialization (type C)
Harder to influence the course of events at work

Figure 2. Pros and cons of remote working
a) The individual

THE EMPLOYER

Advantages
Higher productivity, less dead time, lower wage costs (type C) and
Lower costs for premises, heating etc (type C)
Easier to recruit carce staff (type B)
Buffering: coping with seasonal rushes without taking on more staff (type C)
Lower social and tax costs (type F)

Disadvantages
Hard to supervise and check up
Greater demands on organization

management
Clients may be negative
Extra costs (when staff need two work places or with part time workers)
Technical problems of transporting or transferring work

Figure 2. Pros and cons of remote working
b) The employer

THE UNIONS

Advantages
Shortage of local jobs

Disadvantages
Hard to reach remote workers
Hard to represent remote members
Problems with Swedish law, often not applicable/enforceable away from the work place

Figure 2. Pros and cons of remote working
c) The unions

SOCIETY AT LARGE (POLITICIANS)

Advantages
Means of achieving regional balance
Means to create jobs for the handi-
capped
Reduced energy consumption for transport, reduced pollution
Less pressure on large cities

Disadvantages
Principles concerning working life, equality, working environment

Figure 2. Pros and cons of remote working
       d) Society at large (politicians)

Figure 2 would seem to illustrate very clearly just why the predicted revolution (7) is so long in coming. There is an inherent conflict of interest between employer and worker: those forms of remote working which are profitable for the employer are generally type C, that is, those which carry the most disadvantages for the employee - partly, but not principally, due to current legislation.

There are two exceptions to this apparent rule. One is the need of employers to recruit, retain and pamper scarce personnel (type B), and the other is the case of the employee who is prepared to accept the disadvantages of type C home working (10) in order to achieve other benefits. For example, in a society with poor or no arrangements for child care, a parent may choose to work at home as being marginally preferable to having no income at all, despite all the difficulties.

A ray of hope could still be offered by the corner office. If it were to be established and made viable with the help of local employers, self-employed people and elite groups of type B remote workers, it could perhaps provide a reasonable-cost alternative for remote working of other kinds. Perhaps the scenario in our original paper (1) was not entirely unrealistic. But so far we see very few signs of actual commitment from any influential group. By the same token type C home working seems to have gained no ground in Sweden and will probably not do so as long as the country remains committed to present policies e g as regards child care. What we have seen over the past five years has been more words than action. And in the light of the above, we should perhaps be glad - despite the undoubted potential for benefits.

REFERENCES:

(1) Hedberg, B. and Mehlmann, M., Computer power to the People (Swedish Centre for Working Life, Stockholm, 1981)
(2) Hedberg, B. and Mehlmann, M., Computer power to the People: Computer resource centres or home terminals? Behaviour and Information Technology (1984, volume 3, no 3, pp 235-248)
(3) The Gospel according to St. Matthew, XXV.21.
(4) Engström, M-G., Paavonen, H. and Sahlberg, B., Tomorrow's Work in Today's Society: a full-scale experiment in Sweden (Swedish Council for Building Research, Stockholm, 1986)
(5) Engström, M-G., Paavonen, H. and Sahlberg, B., Grannskap 90: närarbete på distans i informationssamhället (Teldok rapport 16, Stockholm, September 1985)
(6) Grönberg, E-B., Ljungberg, C., Paavonen, H., Engström, M-G. and Sahlberg, B., Förvärvsarbete i bostaden - Förstudie (Nordiska institutet för samhällsplanering, Stockholm, Meddelande 1984:8)
(7) Toffler, A., The Third Wave (Pan Books/Collins, 1980)
(8) Naisbitt, J., Megatrends (Futura/Macdonald & Co, London, 1984)
(9) SAF:s Allmänna Grupp, Distansarbete (undated)
(10) Elling, M. and Parmsund, M., Långt borta och nära (Swedish Centre for Working Life, Stockholm, 1982)
(11) Elling, M., På tröskeln till ett nytt liv? (Swedish Centre for Working Life, 1984)
(12) Hedberg, B., Elling, M., Jönsson, S., Köhler, H., Mehlmann, M. and Werngren, C., Kejsarens nya kontor (Liber, Malmö, 1987)

3

ORGANIZATIONAL AND TECHNICAL ASPECTS

THE DILEMMA OF TELEWORK: TECHNOLOGY VS. TRADITION

Gil E. Gordon *

1. INTRODUCTION

After approximately 100 years of efforts to centralize the workplace since the Industrial Revolution, we are now seeing serious attempts to decentralize work - especially office work - as a way to respond to corporate pressures and employee preferences. We are amazed at the range of technology available to help make this change, there is growing interest among employees to work at home or elsewhere off site, and there is a growing list of rational reasons why it should happen. However, it is safe to say that telework has not lived up to its expectations. While this is disappointing to some, I believe it is an outcome that can readily be explained if not welcomed.

In this paper I will examine five issues:

- What has accounted for the growth of telework to date?
- Why haven't we seen more growth in telework?
- What is the role of technology and just how important is it?
- In what ways has the government helped or hindered its growth?
- What are some of the likely future scenarios for telework?

1.1. A Definition

Before addressing these issues, a few points of explanation are in order. The focus of this paper is on what I refer to as "telecommuting". The people who are telecommuters are employees of organizations (and thus are not self-employed), and routinely spend between two and four days a week working away from the office. In most cases today, this time spent away from the office is spent at home, though that is far from the only kind of remote work location. But since the home is so common as a remote work site, I will refer to the "home" as the predominant alternate work location.

The time spent at home is _in addition to_, not in place of, time spent in the office. Thus, this is not the kind of casual work done at home during the evening, nor is it work at home done by people who have always worked out of their homes such as salespeople or writers. I am concerned only with the substitution of a remote work location for people who would otherwise commute every day to the central office location.

This is a narrow definition yet an important one. While many people work at home for many different reasons, many of the important policy questions only arise, I believe, when we look at the routine, ongoing use of the home as a remote work loca-

---

* Gil E. Gordon is President of Gil Gordon Associates, 10 Donner Court, Monmouth Junction, New Jersey 08852 (USA). Telephone number 201-329-2266.

tion. For example, questions of adequate remote supervision
rarely if ever arise when considering an employee who spends
a few hours working at home in the evening after he or she has
completed a normal work day in the office.

2.   WHAT HAS ACCOUNTED FOR TELEWORK'S GROWTH TO DATE?

     My current estimate of the number of telecommuters – defined
as above – in the U.S. today is approximately 10,000 people.
This is a large number when viewed as a sign of rate of adoption
of an innovative work practice, but a small number when compared
with some rather high projections of only a few years ago. As a
percentage of the U.S. labor force, it is a tiny proportion, well
under 1% – and would still be at that level even if my estimate
was multiplied by a factor of five.

     In the next section I will examine why the number is not
much higher. For now, let us look at the reasons why even this
small number of employees are working at home. Here are eight
reasons why U.S. employers have begun using telecommuters; the
items at the top of the list account for more interest than those
near the end, but this is not an absolute ranking by importance:

1. IMPROVED RECRUITING – Any time there is an imbalance between
demand and supply of particular talents, employers are likely to
experiment with new ways to attract qualified applicants, and set
themselves apart from other employers in the labor market.

     Telecommuting has been used in this way, especially in the
data processing field. For example, programmers may find it
appealing to be able to do some or much of their work at home and
will seek out employers who offer this privilege. There is a
potential problem with this approach, since it can create resent-
ment among other employees who are not given this option.

(The cartoons appearing in this paper originally appeared in the
"HOMER" feature of the monthly newsletter **TELECOMMUTING REVIEW**,
edited and published by the author.)

     The other recruiting-related advantage is to attract people
who otherwise could not or would not be able to work in a normal
office setting. Whether this is due to family needs, health, or
simply personal preference, there are many qualified workers who

are never in the labor pool if an employer only seeks the people who can and are willing to work in a traditional office.

2. IMPROVED RETENTION - This is the opposite of the previous point. Just as it is important to <u>find</u> good people, it is important to <u>keep</u> them once they've been trained. Most employers don't keep accurate records of the costs of losing good people and then finding and retraining a replacement. I have seen estimates ranging from $30,000 to over $100,000 (US) for replacing a professional-level worker.

Telecommuting allows employers to retain a trained, trusted employee who might otherwise have to leave the workforce. It will not work for every departing employee but even when used selectively it offers a benefit.

3. CURIOSITY TO EXPERIMENT - In the early 1980's some of the telecommuting trials resulted from nothing more than someone's curiosity to see how well the innovation would work. This is especially true as personal computers came into the business world, thus making computing power very portable.

This curiosity is a mixed blessing, in my view. It is certainly important to have someone act as "sponsor" for trying an innovation like this, and American corporations have often been criticized as being lacking in innovation. However, if the <u>only</u> reason for trying telecommuting is pure curiosity,, it is not likely to last as long or work as well.

In organizations where experimentation has been the driving force, it has been hard to continue the program once the employer sees that it is in fact workable. Also, once the sponsor (often a person with enough rank or power to make sure the innovation gets tested) changes jobs, the project generally ends.

There must be business-related benefits in mind for telecommuting to last. Ideally, telecommuting must be seen as a good solution to a business problem facing the organization; this helps maintain interest and overcome initial resistance.

4. SPACE SAVINGS - Businesses that are growing quickly and need more space, and/or want to pay less for office space, have found telecommuting of interest. Since it costs between $1500 and $6000 (US) annually for office space and related services for one employee (depending on the city and the type of space), there is good reason to look at ways to cut office space needs.

While the cost-saving associated with requiring less office space is perhaps the most tangible benefit from telecommuting, it has not proven to be a major factor. An architect whose firm designs large office buildings for corporate clients told me that when a client planning to spend millions of dollars on a new office, he really isn't interested in saving what would be a very small percentage of that cost by using telecommuting. This is perhaps a narrow-minded but realistic appraisal.

This may change over time as more and more jobs become suited for telecommuting, and the percentage of the workforce involved gets higher. If up to one-third of the workforce can work at home several days a week, the savings are no longer insignificant. I believe that large employers will begin to include telecommuting in their long-range facilities planning process. They can consciously "under-build" - that is, they can plan to build or lease less space than they need, and make up the difference with telecommuting.

This perspective is more likely to get the attention of top management that is involved in the financial implications of facilities planning five and ten years into the future.

5. HIRING THE DISABLED EMPLOYEE - Some telecommuting projects have been designed from the start to address the needs of the handicapped or disabled. This has been especially true for companies that have a strong sense of social responsibility and want to provide jobs for people who otherwise would be unemployable.

There are also several private and government-sponsored agencies that act as intermediaries, working to find and train the disabled and then placing them for employment. These plans have generally been very successful but the numbers are, once again, quite small. There are some observers who say that telecommuting should not be stressed for the disabled, because this group has been kept out of the main office environment for too many years. Every effort should be made, they say, to bring the handicapped into the office and not keep them at home where they have been for too long already.

6. INCREASED PRODUCTIVITY - It has only been since the early 1980's that U.S. employers have begun to be widely concerned about the productivity of office workers. Before that time, productivity was a concept thought to apply only to factory workers. But with the growing percentage of employees - and salary dollars - in the office today, the need to understand and improve office worker productivity is also growing.

One of the most consistent findings from telecommuting programs has been the increase in productivity. Gains in the range of 15% to 25% have been typical. However, these gains must be viewed cautiously since office-worker productivity measurement is far from an exact science. The gains that have been reported are the result of many factors, including:

- MORE HOURS WORKED PER DAY - In many cases the people convert some or all of their travel time to work time.

- MORE WORK DONE PER HOUR - Once they are away from the distractions and interruptions common in many offices, telecommuters find they can do more work simply because they can concentrate better. Of course, this depends on the setting in the home; without appropriate work space and separation from family activities, the person's productivity may very well go down.

- BETTER MATCHING OF WORK HOURS AND BIOLOGICAL CLOCKS - The typical 9-to-5 work schedule is more a convenience to the employer than the employees. Some people are most productive in the very early morning hours and others work best well into the night. Telecommuting has allowed people to produce more and better work by allowing them to work at a time that is their personal peak period, instead of according to the employer's timeclock.

- LESS INCIDENTAL ABSENCE - There are many times when employees have to attend to personal business and therefore lose part or all of the work day, even if the personal business takes only a few hours. Similarly, there are times when an employee wakes up feeling ill and tells the employer that he/she will be out sick. If the person improves by mid-morning, it is rare that he/she would come into the office for the remainder of the day.

In both of these examples, telecommuters are able to salvage part of the day by working at least a few hours at home. These kinds of incidental absences would seem to be minor, but their cumulative effect can be significant.

- FASTER PROCESSING TIME - Telecommuters who use computer terminals or personal computers linked to a main computer at the office find that they can often get more work done later in the afternoon or into the evening. When fewer employees are using the same main computer on a "timesharing" system, each person can get faster responses and will produce more work. This, along with the "biological clock" explanation noted earlier, is a reason why many telecommuters shift their work hours away from the traditional schedule.

In addition, this schedule change offers other benefits to the employer. When a portion of the computer's workload can be shifted to the evening hours, it often results in a more balanced utilization of that computer. That is, a more constant percentage of its capacity is used around the clock, instead of it being used most heavily during the day and much less so at night. This workload balancing means that additional computer capacity might not be needed as soon as had been planned, resulting in large savings.

In summary, the productivity gains from telecommuting have been among its most notable benefits. One company in New York has sixty people at home doing as much work as eighty people did in the office, without increasing the error rate. This labor savings (and the office space savings) cannot be overlooked.

7. EMPLOYEE INQUIRIES - There are two schools of thought about how telecommuting gets started in a company. One says that it is management's initiative, taking a "top-down" approach that takes advantage of the cost savings and other gains noted here. The other says that employers will respond to "bottom-up" pressure from employees who want to become telecommuters. The latter

accounts for some portion of telecommuting today but less than might be imagined.

There is one major exception to the last statement, which is the growing number of professional-level workers who have purchased personal computers for their homes and use them to do after-hours work. They commonly transport diskettes to and from the office, and in some cases telecommunicate with computers in the office.

This after-hours work does in some cases lead to work at home in place of time in the office. The employee begins to ask, "why should I take the trouble to get dressed in my business suit and commute into the office, just so I can sit at a desk and work at my personal computer - just as I'm doing now in the comfort of my home?"

It is my belief that many more employees are _interested in_ telecommuting than are _qualified_ to do it well. Employers who have responded to this kind of upward pressure sometimes find that people want to work at home simply because they like the more casual work environment and the convenience. These are important benefits, but they ignore the need for more discipline and self-control. Therefore, while employee interest has helped encourage some employers to consider telecommuting, it has not been sufficient to sustain it over the long term.

8. IMPROVED CUSTOMER SERVICE - Finally, the last main reason for telecommuting's growth to date is the business benefit of providing extended customer service hours. One organization in Chicago, for example, provides 24-hour toll-free telephone service to its customers. The company was having trouble recruiting people to come to the office to work in the middle of the night. As a result, customers were not always able to have their phone calls answered during the night. Also, the company had to pay for keeping the building open (lights, heating, and so on) for a very small number of workers.

The solution was to hire four people to work at home from 11 p.m. until 7 a.m. to answer phones. The company was able to find employees who enjoyed those late hours - especially when they could work in the comfort of home. The result was that customer service was maintained, and the costs of keeping the building open during the night hours were eliminated.

While this has not been a major factor behind telecommuting in the past, I believe it will grow quite quickly. Because there are so many couples where both husband and wife work, and so many single-parent families, there is less and less time to go shopping in stores. This has led to much growth in catalog shopping, especially involving the toll-free telephone ordering. Customers can pick up the phone at any hour of day or night and place an order - and there is no reason why the person processing the order at the other end cannot be sitting in the home instead of the office.

In summary, then, we have seen eight reasons why remote work as I have defined it has grown to the level it is at today. One reason is notable absent from this list is growth in technology to support telecommuting, particularly the personal computer. This was briefly mentioned in the "employee inquiries" section but nowhere else. There is no reason for me to believe that more personal computers to date has meant more telecommuting - but I think this will be changing soon, as will be discussed in a later section.

3. ORGANIZATIONAL LETHARGY: WHY HAVEN'T WE SEEN MORE GROWTH IN TELECOMMUTING TO DATE?

If we could believe what some of the futurists were saying back in 1981 and 1982, we would have expected to see up to 10% of the office workforce spending several days a week at home or elsewhere off-site by this time in early 1987. We are nowhere near that number, even allowing for a large estimating error.

The question must be asked, "why not?" If the concept is technologically possible, provides good business benefits, and appeals to many employees, why isn't it more widespread? I believe the answer lies in the six reasons described below.

First, though, please note that this lack of progress should not be seen as a negative. While a more rapid adoption of this concept would be gratifying in some respects, it might indicate a "fad" or trend that, like other trends, has a short life. It is very reasonable for large employers to take a long time to study and become accustomed to this concept. I would rather see this happen than to see more rapid adoption that might end just as quickly because it was not analyzed properly at the outset.

3.1. The Myths Of Telecommuting

In its short history, telecommuting has been surrounded by a number of myths that appear to limit its use because they cause employers to back away from the idea.

I have identified four such myths:

Myth #1: "Telecommuting is mostly for women who are at home taking care of young children."

This is one of the most common perceptions about working at home - that you can easily work while caring for children. With few exceptions, nothing could be further from the truth.

The vast majority of telecommuters who are working on a full-time work week (or anything close to full time) have made other child care arrangements. The myth is that the telecommuter can sit at home with a baby bottle in one hand and a keyboard in the other. While this might be the wish of some managers, it is virtually impossible to do. Either the job, the child care, or both will suffer in the process.

Telecommuting does allow some people to better balance their work and family duties, in the case where they are working part-time (perhaps 20 hours a week) and have other child-care help available. But telecommuting is not, unfortunately, a good way to end the severe lack of child care resources in the U.S.

But the perception that it will work this way limits its broader application to men and to employees who are not parents.

It has been noted that telecommuting fits well with the traditional beliefs of some male managers that women really should be at home with the children and not in the business world. If that is the case, the companies for which those managers work are missing an opportunity to take advantage of telecommuting for many other kinds of employees.

Myth #2:   "Telecommuting means working at home."

As was stated in the introduction, this chapter equates telework with home work simply because that is where most remote work has been done to date. While this is true, there are other telework applications today in satellite offices, branch offices, and even inside the walls of prisons. We are dealing with the idea of decentralizing the work - the home is one place in which this will occur but not the only one.

A strong argument can be made that the use of satellite offices or neighborhood work centers will grow as fast or faster than home work will. The reason is simple: a major reason why managers resist work-at-home options for their employees is the perceived lack of supervision. This problem does not exist in a satellite office, and the ability to have on-site supervision calms many of the fears of reluctant managers. Also, it will be more economical to provide certain equipment (such as copiers or laser printers) in a satellite office (where it will serve many people) than in the home (where it will serve only one person).

However, to the extent that a manager believes that telework means only home work, that manager will resist the use of telework if he/she believes it cannot be properly supervised.

Myth #3:   "Telecommuting is just for computer programmers and other data processing jobs."

The high-technology image of telecommuting is sometimes its own worst enemy. There is no doubt that many telecommuters work in data processing jobs, but I believe this reflects the nature of the labor market imbalances more than anything else. What many managers do not realize is the large number of telecommuting jobs today involving only paper-and-pencil work, and also perhaps a standard telephone.

If organizations consider only those jobs where people sit at computers or terminals when considering telecommuting, they are taking a very narrow view. We know there are certain criteria that describe jobs well-suited to telecommuting, and these jobs can be at any level and in any department - and with widely varying amounts of technology involved.

Myth #4: "Telecommuting means being away from the office full time - five days a week at home."

We have come to realize that the flexibility of telecommuting is its best asset. While some telecommuters do work at home five days a week, this is not at all common. This schedule is generally not practical either for them or the employer.

The remote workers have to come into the office on a regular basis to attend meetings, make presentations, and socialize with co-workers. While it might be possible for them to do their complete jobs at home full-time, this is likely to lead to problems over time. Managers who think that five days at home is the rule are understandably hesitant to adopt telecommuting.

In fact, one of the best ways to get a program started is to have the telecommuters work at home only one or two days a week for the first month or two. This very gradual implementation makes it much easier for everyone to accommodate the change, and gives the manager a sense of security about the transition. This schedule might lead up to a maximum of four days a week at home, but only after everyone is satisfied that the work can still be done effectively.

3.2. The 100-Year-Old Tradition

Old habits last a long time, and the 100-year-old tradition of going to a central workplace is an old habit. The phrase "going to work" is part of the language and there is an entire set of customs and routines based on that daily two-way journey.

What is interesting is that we have lost sight of the cumulative effects and costs of that daily habit. According to survey data from the Hertz Corp., the car-rental company, in 1985 106.4 million people commuted an average of 24.4 miles round trip to work each day in the U.S. Commuting (by any method) in 1985 cost $1,355 (US) per person, and 69% of the U.S. workforce commuted to work in their own cars.

If we expand on these numbers and assume that the average 24.4 mile round-trip commute takes forty minutes per day, that adds up to 8,119 YEARS of time those 106.4 million people spend commuting every day. The total cost of commuting, using the $1,355 average, was $144,172,000,000 (US) in 1985. These are somewhat astounding numbers, and they do not even include the various subsidies employers provide for commuting (such as parking lots or reimbursement for parking garage fees).

Despite this, the weight of tradition still prevails. The unfortunate – but understandable – irony is that if a company with 20,000 employees starts a telecommuting pilot program with 10 employees, the company believes it has accomplished something noteworthy. In fact, they have – even though it involves only 10 employees, the pilot program is a tangible sign that the company is willing to begin challenging or even breaking that tradition.

But we should never underestimate the difficulty of changing habits in large organizations. Big companies became big because they made slow, deliberate progress, and not because they took every opportunity to innovate and try new methods. Whether we like it or not, this is a common fact of organizational life.

3.3. The Fear Of Loss Of Control

This is perhaps the biggest reason why telecommuting is showing slower progress than some had expected. The philosophy of management that is prevalent in many organizations goes back to a legacy of factory supervision, in which close observation of direct labor was common. Even though we have progressed from the factory to the office, many of our supervisory methods have not changed much.

Many managers think that to supervise means to observe; in fact, the Latin basis of supervise is "super" (over) + "videre" (to see). When managers cannot "oversee" their workers they become uncomfortable - one common question raised by managers about telecommuting is, "How can I manage someone I can't see?"

The answer is to make the distinction between <u>observing activity</u> and <u>managing results</u>. The most successful managers of telecommuters are those who concentrate on the "deliverables", i.e., what are the end-products that my employee must deliver to me by a certain date and according to certain standards.

It is most interesting that managers of telecommuters almost always say that the process of having to manage from a distance makes them better managers of other employees who remain in the office. This is because the manager is now forced to use some of the good management practices that have been taught for years (such as setting goals and monitoring progress). These are often ignored because the in-office manager has the luxury of frequent, close contact with the employees.

It has been suggested that it is a sign of the manager's own insecurity with his/her managerial skills that leads to this reluctance to manage from a distance. I don't know if this is true, but have observed that it seems to be the more competent managers who are among the first to embrace the idea of telecommuting and can see its benefits.

Another way managers express their reluctance to supervise telecommuters is by asking, "How will I be able to tell when my telecommuter is doing a full day's work at home?" This is another example of the very understandable fear of loss of control felt by the manager. My reply to this question is, "How can you tell your employee is doing a full day's work <u>in the office</u>?"

This takes the manager by surprise, and usually shows that the manager in fact does not have very good ways to describe and measure progress toward the "deliverables", and instead manages by observing activity. It is presumed that high activity levels will lead to the desired results, but this is far from assured.

As frustrating as these attitudes are to those who advocate telecommuting, it is important that we do not ridicule or criticize the manager who fears the loss of control. In some cases the manager is relying on methods he/she has found to be effective in the past; in other cases that manager is simply responding to the expectations for how managers should manage in that organization. These attitudes and beliefs about supervision will not change overnight, yet they must begin to change before telecommuting becomes more widespread.

3.4. The Role Of Job Evaluation Systems

In the U.S. and elsewhere the pay level of a job is often determined by the nature of responsibilities, scope of decision-making, and other factors. Most times, these "job evaluation" systems generally do well at measuring the differences in value of different jobs, to insure that managers whose jobs add more value to the organization are paid more.

Interestingly, one of the determinants of a job's value is the size of the organization managed by the manager, and in some cases even the size of the manager's budget. If we put ourselves in the position of a manager, we can see how telecommuting can be threatening if it leads to reductions in the size of the organization. Specifically, the manager might ask, "If some of my people no longer work in the office, and if that means I will be responsible for a smaller work area, will that mean my job will be evaluated lower - and therefore will be paid less?"

I am not aware of cases where this has happened, but can easily see how a manager could use that kind of logic as a good reason why not to allow telecommuting in his/her group. Very few managers would be willing to try an innovation if they thought there was any chance it would lead to a reduction in salary.

There is, incidentally, a simple way to counteract this concern. If management would be willing to share a portion of the savings that resulted from telecommuting, I think this would more than make up for the concerns. The practice is now starting to be used more widely; this is an example of how to create an incentive for something like telecommuting by providing a reward for implementing the innovation.

3.5. The Effects Of Worsened Business Conditions

When competition increases and profits are dropping, the climate for risk-taking is generally poor. Worldwide business conditions are more competitive and, in the U.S., companies are eliminating employees and taking other steps to control costs.

This trend has been developing over the last five years and there are signs that it will keep growing. With so many managers having to do as much or more work with fewer people and other resources, they are often not willing to pursue something like telecommuting. They see it as a distraction, in the sense that their time for any new project is so limited.

This very situation could turn out to be a good reason why telecommuting __will__ grow. If (and it is a big "if") managers can see the link between telecommuting and cost-cutting, improved retention of valuable employees, and improved customer service, they will then see telecommuting as a solution to business problems. This has been the case in some organizations where telecommuting has worked the best.

3.6. The Fear Of Employee Lawsuits

This last factor is minor but worth mentioning. In the U.S. there have been more lawsuits by employees against employers, as a means of settling employment disputes. This has made many organizations somewhat hesitant to introduce new programs where there are legal uncertainties (as does telecommuting) for fear of opening the door to new lawsuits.

This could affect telecommuting's growth because of the <u>perceived</u> uncertainty about several aspects of remote work, such as levels of pay, amount of benefits to be paid, and liability for injuries or accidents occurring in the home while the person is working there. To my knowledge, there has been only one lawsuit to date in these areas but some employers are still wary.

That one lawsuit is a very important one, at least in the U.S. It involves a group of eight telecommuters working for Cal-Western Life Insurance in Sacramento, California. They were among a group of almost thirty telecommuters who had been working at home for over two years doing insurance claims processing. The suit involved the issue of whether or not they were actually employees of the company, instead of so-called "independent contractors" who received no benefits and were paid strictly on a per-unit-of-work basis. (As of the time when this paper was written the lawsuit still had not been heard in court.)

The definition of employee status (that is, employee vs. independent contractor) is a critical issue in telecommuting. One way the problem arises when an employer has a person working as an employee in the office who then becomes a telecommuter working at home. If the employer chooses to convert the person to independent contractor status - even with the person's consent - this may be illegal under various state and federal laws. Such a conversion is very attractive to employers because it lets them eliminate insurance benefits and other "overhead" expenses that add between 30% and 40% to payroll costs.

I have taken the position that it is not only illegal in most cases to carry telecommuters as independent contractors, but it makes poor business sense to do so. Even though there are significant costs savings, these must be compared with the long-term effects of maintaining a non-employee relationship in which loyalty to the job and company often suffer. Also, when telecommuting is implemented properly, there are enough built-in savings that it is somewhat greedy to go for the payroll savings as well.

The Cal-Western lawsuit has probably made some companies think twice about getting started with telecommuting, but it also has made others reconsider their plans that were in progress. This is positive, since in some cases the companies decided to keep the telecommuters as employees instead of shifting to independent contractor status. This decision benefits both parties.

Summarizing this section, we see there have been a number of complex reasons why telecommuting has not grown to the extent that was expected. Some of predictions simply did not take the factors that have been discussed into account. There is no doubt that the number of jobs that <u>can</u> be done with telecommuting is in the millions, but the number that <u>will</u> be done, especially in the short-term, is much smaller for what I think are very understandable reasons. This does not mean that the growth will not happen - it will just take longer than expected.

4.   THE ROLE OF TECHNOLOGY: DRIVING FORCE, CATALYST, OR OBSTACLE?

It is a little difficult to determine exactly what role is played by technology in the growth and spread of telecommuting. I believe there are three possible explanations:

- It is a <u>driving force</u> to the extent that the spread of personal computers (PCs) and developments in telecommunications almost make it imperative that we rethink how and where office work should be done. There is an analogy in the growth of the automobile; once the population had easy access to a very personal kind of transportation, the old assumptions about housing, shopping, and recreation had to be challenged.

- It is a <u>catalyst</u> to the extent that it lets us selectively reorganize and decentralize the office, but within the limits that the basic work procedures and workflow stay the same.

- It is an <u>obstacle</u> to the extent that we are overwhelmed by the choices we have in computing and telecommunications, and the idea of large numbers of remote workers adds a level of complexity that many data processing managers will not want to consider.

In reality, there is a little of all three happening. But I believe it is a mistake to assume that there is or will be a direct, causal link between the spread of PCs and the spread of telecommuting.

This kind of technological determinism ignores two factors. First, there is the whole list of obstacles listed in the last section that are essentially technology-free in nature. Second, there have been and will be many, many excellent telecommuting applications that rely very little on technology. These are paper-and-pencil jobs, and perhaps the standard telephone is the highest level of technology that is used.

But we cannot ignore the spread of PCs as a factor. We see companies in the U.S. moving towards a ratio of one PC for every professional-level employee in the next few years, for example. The more that the job is "contained" in the PC, the more that job is made portable to the home or elsewhere. Also, as was noted before the number of employees who purchase PCs for use at home and then begin using them for office work is another example of how remote work is driven by PC growth.

There are some examples of how technology can facilitate remote work by changing some of the assumptions about where certain kinds of work can be done. Here are four chosen from very different fields:

1. PROJECT VICTORIA - The Pacific Bell telephone company in California conducted an experiment in 1986 that showed how one standard residential telephone line could be transformed into seven digital channels. Two are for voice - one for medium-

speed and four for low-speed data transmission. Project Victoria used a proprietary multiplexing technology to divide the phone line, and the test was conducted for four months with 200 participants in Danville, California.

The significance of this test cannot be stressed enough. This is the first time that such a wide range of data and voice transmission could be done using one unmodified residential phone line, simply by adding the appropriate "black box" at the central office and within the home. Once this technology is brought to market, employers will be able to set up remote jobs that require high-volume data transmission, without incurring the costs of special "dedicated" phone lines.

As an aside, some of the same benefits will occur as ISDN technology is implemented. ISDN, however, will probably not be in wide use for a number of years.

2. SMALLER FACSIMILE EQUIPMENT - We are now seeing a number of small, easily portable facsimile units with full Group III compatibility and purchase or lease costs that are well within the range of most employers. In some cases these are integrated with telephones, which is appealing for telecommuters because it takes less space on the desktop.

The advantage here is that jobs involving graphics are now more "portable" to the home since images can be transmitted as readily as text. Without this equipment, the telecommuter in a job like a commercial artist or consumer marketing manager would have to rely on frequent trips to the office, courier service, or more costly PC-based video-to-digital systems.

3. PORTABLE MICROFICHE READERS - Next, we have the rather old technology of microfiche coming into play in telecommuting since portable microfiche readers much smaller than a briefcase are available for about $150 (US). This allows a job involving microfiche to be decentralized, since the reader can move wherever the person is. Also, it may be a good solution for jobs that rely heavily on paper files (such as accounting or customer service) where the microfiche images of documents can be transported more easily than the documents themselves can.

4. REMOTE ACCESS/REMOTE CONTROL SOFTWARE - Several interesting PC software packages came on the market in 1985 and 1986, all designed to enable a PC user to take control over a PC at another location. Examples of these are Carbon Copy (tm), Remote (tm), pcANYWHERE (tm), and CloseUp (tm).

There are two ways these programs work. First, they allow two PCs to work together so whatever is seen on one screen is seen on the other. One application is training or problem-solving; person A is trying to learn to use a word processing package and person B (at another location) is acting as tutor. Person B can type a command - or even a comment - on his/her keyboard and have the result show up on Person A's screen.

The other application is to run a program on or gain access to a PC at another location. If a telecommuter is at home and wants to retrieve information from the hard disk of a PC in the office, or perhaps run an applications program on the office PC because it is a faster machine, this can all be done with these packages <u>without</u> anyone having to touch the PC in the office.

The benefits of these packages to telecommuters are clear: there is no need to be physically located with the person or PC with which they interact. All of the contact can be done over regular telephone lines as needed.

4.1. Connectivity and Security: Two Special Concerns

If these kinds of technological innovations are readily available, how can it be said that technology is an obstacle to telecommuting? The problem is not with the individual "pieces" of technology, but with how they are linked together. Advances in computing have not been matched with advances in telecommunications, though the latter is quickly gaining ground. Also, in the rush to buy PCs over the last few years, many companies bought them with little or no thought to standardization and ease of connection.

As a result, one major theme in data processing circles today is "connectivity," which is a fancy word for the process of trying to link equipment together that in some cases was never intended to be linked. This occurs from PC to PC and from PCs to minicomputers or mainframes. It creates headaches for data processing and telecommunications managers who are asked to make these magical connections. What makes matters worse is that the people doing the asking are non-technical senior executives who have been convinced by advertising that everything and anything can be done "just by pushing a few buttons."

Telecommuting is in the midst of this connectivity arena, and in some companies there simply has not been enough time to sort out all the combinations and possibilities. Depending on the complexity of the combination and the type of data communications intended, it becomes either an easy task or a technical nightmare to make it work.

It is encouraging to note that there have been very, very few reported cases where a company wanted to implement a remote work program and was unable to because of technical problems. In some cases the telecommuters may not have been able to perform all the tasks they wanted to with their home PCs (compared to what they could normally do in the office) but were still able to do their jobs.

If connectivity is one technical concern, data security is another. With the publicity given the "hackers" who seek to break into computer systems, there is concern that the idea of telecommuting in some ways invites problems to occur. Whenever a dial-up system is used to allow for remote access, there is an added risk. To my knowledge there has been no reported case of computer crime caused by a telecommuter, although companies do not always publicize their security problems.

Security is an interesting part of the technology discussion because it shows how talking about technology alone is only part of the problem. Security experts admit that a determined and resourceful criminal can break into any computer system; however, the security devices on the market today are adequate to prevent all but the most determined "hackers". But there is often a big gap between the adequacy of the security devices on the market and the computer security practices in large companies, and that is where the problems begin.

Specifically, the companies will make one or both of these two mistakes. First, they will not take advantage of the various software and hardware available to provide good security. For some unexplainable reason, the company may believe no one would want to bother trying to get into its computers or believes its current minimal level of protection is enough. Second, even if good security software and hardware is in use, the company may not routinely monitor its use and take advantage of the various audit and reporting features to detect attempted break-ins.

Therefore, when we talk about the security risks of remote work we have to think about the <u>incremental</u> risks; that is, what if any <u>additional</u> risks are there. If a company has lax security procedures to begin with, it is hard to claim that telecommuting should not be implemented because of its presumed security risks.

Also, there are other contradictions between what companies say and what they do in the security field. I have worked with large, sophisticated companies that claim to be very concerned about the threat of security problems from telecommuting. Yet in these same companies I have seen terminals in open office areas that have a piece of paper with the current password taped to the terminal, in plain view of anyone walking past.

I also ask one simple question: can your employees walk out of the office with a diskette in a briefcase? If the answer is yes, that is a good sign of possible security problems - no matter how much the company <u>claims</u> to have strict security.

This is not to minimize the very real possibility that some telecommuters might cause security problems. Also, in employers such as banks or hospitals where confidentiality of the data is essential, telecommuting may never be appropriate no matter how many security systems are in use. Management might not ever be convinced that the risk is controlled, and (correctly) will avoid telecommuting or limit it to non-sensitive positions. My point is simply that we cannot assume there is an <u>automatic</u> security threat whenever telecommuting is introduced; we have to look at the entire security picture in the company first.

In summary, the role of technology in telecommuting in the U.S. is far from clear. The PC revolution is undoubtedly a part of telecommuting's growth and appeal, but by itself it will not create an explosion of new telecommuting applications. We are fortunate to have the wide range of hardware, software, and telecommunications available from which we can choose, even though the choices can be overwhelming at times. Technology in its many forms is a necessary part of telecommu-ting's future but is not sufficient in itself to fully decentralize the workplace.

5. THE ROLE OF THE GOVERNMENT: REGULATION, PROTECTION, INCENTIVES, AND OBSTACLES

Telecommuting in the U.S. is affected by a mixture of federal and state laws, many of which unfortunately seem to be inconsistent with each other. In addition, there is a wide range of laws that were never intended to deal with telecommuting but that still must be reviewed for possible applicability. While there are no federal laws today that directly limit telecommuting in any way, there are some state labor laws affecting work-at-home that indirectly can make it difficult.

In this section I will discuss three issues:

- Employment-related regulations, such as those affecting employment status and worker's compensation.

- Other indirectly-related regulations, such as zoning and income tax provisions.

- Government incentives and disincentives to telecommuting.

5.1 Employment-Related Regulations

The regulatory issue of most concern to telecommuting is labor laws affecting the terms and conditions of employment. This was addressed earlier in the Cal-Western lawsuit discussion. The question is whether telecommuters are, should be, or want to be classified as employees or independent contractors. On one side of the issue are some of the labor unions who, at minimum, want to insure that remote workers receive all the rights and benefits they would get in the office.

However, because the unions believe it is impossible to properly monitor these and other working conditions at remote sites, they have called for a total ban on telecommuting. To date, the federal government has not indicated any willingness to create such as ban. As of late 1986, a proposed change in the U.S. Fair Labor Standards Act was being actively debated. The change would generally remove the remaining restrictions on certain garment industry work done at home.

Whether or not these bans are lifted has no direct effect on telecommuting; however, it does have significant symbolic importance. If they are lifted it would be a signal that the government is highly unlikely to add new restrictions on homeworkers; if they remain, new restrictions become somewhat more likely.

Another problem in the U.S. - and this is generally acknowledged by people on both sides of the debate - is that there is a very weak enforcement mechanism in place to insure that employers comply with existing regulations about employment status. In an era when the Reagan administration has consistently pushed for less government regulation, it seems unlikely that funding will be provided for better enforcement.

So, we are left with a situation where reputable employers continue to treat their telecommuters well (and stay within the bounds of the laws), and where less enlightened employers find ways to "stretch" if not avoid the regulations. This is quite unfortunate for two reasons: first, the victims, if there are

any, are the telecommuters; second, when a few employers treat their people poorly, all employers are likely to pay the price in terms of new laws to correct inequities if they continue.

Most of the concern about inequitable treatment for today's telecommuters centers on clerical-level workers. No one says that people working at home should lose any rights or benefits just because their work location is different. But there is a risk that attempts to mandate equal treatment will penalize the very people who are to be protected.

Why is this so? In the last ten years there has been tremendous growth in what is called "offshore office work" around the world. High-volume data entry and other routine clerical work is now done on contract in countries in the Caribbean, in the Far East, in Ireland, and other locations. The labor rates are so much lower that even with air freight costs to transport source documents and finished work, the total cost still is far below what it costs to do the work here in the U.S. Employers seeking to cut costs are moving more and more work offshore.

It has been suggested that if new laws are enacted restricting the use of telecommuting for clerical work, many employers will simply forget about telecommuting and export that work offshore. Since the Reagan administration is supporting the growth of offshore office work in the Caribbean, through its Caribbean Basin Initiative, it is not unlikely that employers would take their cue from the government and thus avoid the regulations here. I believe this would be unfortunate, because the potential unemployment that would result is a much more serious problem than what we face today with employee-status problems for telecommuters.

There are other regulatory issues in the employment field that affect telecommuting, though these are not as much a factor. For example, the issue of worker's compensation, i.e., payment for work-related injuries or illnesses, is of concern to some employers. The problem with worker's compensation is that each state administers it differently. There is no uniform approach to the definition of what types of injuries are or are not work-related.

This is of particular concern since some kinds of injuries that theoretically can happen to telecommuters are difficult to attribute to the work setting itself. For example, the following question is sometimes asked: "What about the telecommuter who slips on a carpet on the floor on the way into the home office - is that a work-related injury?" Some employers even ask, "What if the telecommuter slips in the bathtub on a day he/she is working at home - is that a work-related injury?"

These situations might seem improbable, and the process of speculating about the "what if's" can even be a little humorous. Unfortunately, it is no laughing matter to employers who have been facing the stream of employee lawsuits noted in an earlier section. Even though these accidents may be unlikely to occur, the employer says "Why should I even take the risk - who needs the aggravation of contending with that kind of problem?" and uses that as justification for staying away from telecommuting.

Fortunately, cases in the worker's compensation field are often decided based on precedents. There are many cases where employees (other than telecommuters) have been injured in their homes, and those cases will be used to help decide any that might arise for telecommuters. Also, employers are finding that the best way to cope with the worker's compensation issue is by using good preventive measures, e.g., training the telecommuters to set up and maintain a safe workplace in the home.

5.2   Other Indirectly-Related Regulations

Two examples of regulations that indirectly affect telecommuting are zoning and income taxes. The zoning issue is like worker's compensation only more complex, since zoning is administered at the level of each city and town. Compounding this is a separate set of restrictions often present in leases for people renting apartments, or in owners' association rules affecting people who live in townhouses or condominiums.

Generally, zoning's intent is to preserve the non-commercial character of homes and neighborhoods. No one would argue with that goal, but problems arise when some rules are written to exclude any and every kind of commercial activity. Many zoning rules were written over fifty years ago when the idea of doing office work at home was virtually unknown. In some cities the same rules that prevent someone from operating a manufacturing business in a home also prevent telecommuters from doing office work at home; Chicago is a good example of this.

The good news is that many cities are revising their zoning laws to accommodate various kinds of work at home, as long as the work does not affect the neighbors or the neighborhood. Since telecommuting creates no outward changes, (e.g., no extra traffic, no noise), it is likely that this zoning obstacle will soon be removed in most cases. But some telecommuters today have to operate almost in secrecy, for fear that neighbors who want to keep the neighborhood "pure" will report them to the authorities.

The second example of indirect effects on telecommuters is the federal income tax structure. For many years it was quite easy to claim the need for an office in the home, and deduct the expenses associated with that office from income before calculating the tax that was due. Over the last six to eight years the government has begun to restrict this home office deduction, to the point today where the only people who can claim it are those who own their own homebased business. Very few telecommuters can legitimately claim this deduction, and they (and their employers) have to demonstrate that working at home was absolutely essential to the employer and not just a convenience to the employee.

5.3 Government Incentives and Disincentives

Looking at government regulations in the U.S. overall, they appear at worst to act as a mild disincentive to telecommuting, but not to the extent that they limit its growth. Therefore, it is interesting to speculate about how the government could change the regulations to provide an incentive, if that was the intent.

I think there are at least three good reasons to consider such a change:

1. The amount of money needed to provide and maintain the transportation systems to get people to and from the office is astronomical. Highways and mass transit systems are all heavily subsidized by the government, which in turn adds to the tax burden. It is imperative that we start looking for ways to move work to the workers, instead of the reverse.

2. We are facing serious unemployment problems in the U.S. as a result of changes in our basic industries (such as steel and autos). I find it much more sensible to begin employing some of these people as telecommuters instead of exporting jobs to off-shore office work operations.

3. We have done a poor job in the U.S. helping families balance work and family life. Although corporate efforts to help provide day-care for employees' children have improved dramatically, there is still a terrible lack of good day-care resources. Telecommuting can be used to help address this problem, subject to the restrictions noted earlier.

How could government create incentives to telecommuting? Here are some possibilities:

1. Provide corporate income tax credits for employers who reduce the number of employees that commute to the central office site.
2. Provide corporate income tax credits and/or reimbursement for start-up and training costs for employers who set up satellite offices in economically depressed areas. For example, a large bank or insurance company in Philadelphia could set up such a remote office in Pittsburgh, a city hard-hit by the decline of steel mills there.

3. Provide real estate tax incentives to employers and/or developers who are willing to create neighborhood work centers by converting unused buildings (such as empty schools). This would not only lessen the commuting burden but would put those vacant buildings onto the real estate tax rolls.

4. Provide real estate tax incentives and/or more favorable zoning rules for employers who can demonstrate they are planning to "underbuild" when constructing a new office, i.e., knowingly planning for less office space than is needed to accommodate future employees, and using telecommuting for the rest.

I realize it is not the right time to talk about providing income tax credits in the U.S., given the size of the federal deficit and the recent tax reform law that raised corporate taxes. However, this is a case of giving up some revenue in the short term to reduce more spending in the long term. However, I am not sure that our legislators will find it politically wise to take that long-term view.

Summarizing this section on the role of the government, we see that telecommuting has probably neither grown nor suffered due to our legislators. The few attempts to stimulate telecommuting among government employees as model projects have not been successful. The best example is in California, where a pilot project for up to 200 telecommuters has been stalled for lack for funding; a smaller version of this pilot may begin in 1987. The best we can say is that the government's actions (or inactions) have not limited telecommuting; it would be more exciting if we could point to actions that _encourage_ telecommuting.

6. SOME LIKELY FUTURE SCENARIOS FOR TELECOMMUTING

The crystal ball into which I gaze when trying to predict the future of telecommuting is very cloudy. Nevertheless, let me try to speculate and make some educated guesses. I am grateful to Dr. Jack Nilles of the University of Southern California - who is the acknowledged "grandfather" of telecommuting - for telling me that there is no such thing as _the_ future. Instead, what we talk about is various futures that might develop. That is a more appropriate starting point when dealing with an innovation like telecommuting.

Research on the adoption of innovations shows there are early adopters and late adopters, and the two are very different. different. Also, the way in which the innovation is implemented varies between the two. This seems to describe what we have seen in the last five years and what we are likely to see in the next five or ten.

6.1.   A Lesson From The PC

As an analogy, look at how the PC came into large companies. It is instructive to remember that companies did not always take the PC for granted as they seem to today. If we can remember back to the ancient history of PCs, all the way back to 1980 and 1981, we see that companies were very adventurous if they bought _one_ Apple II computer. That was a very big step in those days. Typically, they spent a few months experimenting with it, trying to figure out how it worked and whether it had a useful purpose besides playing games or balancing checkbooks.

Once that first hurdle was crossed, the company would then buy perhaps six more Apples, distributing them to departments thought likely to benefit from them. The learning curve was slow at first, the applications were cumbersome, and the time (if not the financial) investment was high. Later, the company bought some more Apples, and eventually some IBM PCs when they came to market, and the microcomputer revolution was underway.

When we see that some companies have literally thousands of PCs today, we have to remember that almost all of them started off with that one lonely Apple II. The same is true of telecommuting; this innovation has to start somewhere, and it is likely to have the same incremental pattern of adoption. The difference is that the PC was seen as an interesting "toy" of sorts, and generally not as the kind of fundamental challenge to traditional business practices as telecommuting is perceived to be.

The companies that were the early adopters of telecommuting have generally expanded their programs, as long as two conditions exist. First, the initial project was set up as a business solution to current or potential business problems and not just as an experiment. Second, the project was planned and managed with at least as much attention to the people issues as to the technical issues. But we are still talking about small numbers; defined as I stated in the introduction, these programs still include at the very most several hundred people per company.

6.2. Some Tentative Predictions

This, then, is the basis of the first part of my prediction. We will continue to see this kind of slow but steady growth in applications within companies where the initial experiences were positive. The growth will occur in two ways: first, we will see more of the same kinds of jobs being done remotely. For example, if a department of fifty programmers had two working at home, that number may increase to eight or ten. Second, we will see new jobs being done at home. Other departments that were watching the initial trials from the sidelines will now be more willing to get involved. Also, technological developments will open up new applications; many of these will be based on improvements in telecommunications and/or PC to mainframe links.

The second part of the prediction deals with the much larger number of employers who were never involved but who may have been following the field as curious observers. Many of them are in larger or more conservative companies that simply are never the first to try something like this but are anxious to be among the leaders in the second generation of adopters. Now that they see the process can be managed and it provides real benefits, they will be more willing to set up pilot projects.

Where will all this lead? I think we will see relatively slow growth for the next two or three years, perhaps, and a much faster growth rate after that. This is based on the assumption that the novelty of the idea will have worn away by then, and it will be accepted as a legitimate, practical business tool by most employers. Even though the proportion of office workers for whom telecommuting is suitable will rise to as much as 50% by 1990, the practical limit is perhaps 10% to 15% of the workforce spending two to four days at home or elsewhere off-site.

I am not convinced that the cultural obstacles to work at home in particular will go away very quickly, if ever. However, the other factor is that we will see parallel growth in remote sites other than the home. One of the strongest forces driving this is the tendency for companies to buy services instead of hire workers, as they try to control their fixed costs. I can envision departments or functional units being severed from the company and then selling back their services, similar to the Rank Xerox experience in England. U.S. employers will probably keep reducing the number of permanent, full-time employees, and remote work in its <u>various</u> forms is one way this can happen.

6.3. Factors That May Increase Or Decrease The Growth Rate

Finally, let us consider a few factors that might dramatically increase or decrease the growth of telecommuting. The first one that could stimulate growth is an external event such as the Arab oil embargo of the mid-1970's. Jack Nilles' original research on telecommuting was done mostly because of that event. I believe the reason his projections did not lead to action then was because the idea of remote computing power was relatively new. Compare that situation with today's wealth of PCs and telecommunications options, and it is easy to see how today's response would be different.

A second factor that could cause a big increase would be a government incentive of the kind listed earlier. Employers are generally very pragmatic, and will be much more likely to implement telecommuting if there is a reward for doing so. This has been shown in programs such as tax credits to hire and train the disabled or the "hard-core" unemployed, for example.

The third positive factor would be the cumulative effect of pressure from employees who want this option. This could be based on family needs, the growth in numbers of homebound aging parents who need to be cared for by their children, or simply a desire to have a more relaxed, convenient work life. I am less certain that this factor alone will create much change unless employers begin to lose large numbers of valued employees because the remote-work option is not available.

On the negative side, the biggest potential obstacles are regulatory in nature. If our legislators passed a series of laws strictly limiting the terms and conditions of employing telecommuters, I think most employers would simply walk away from the idea in all but a few situations. These laws could be motivated by strong pressure from the labor movement, but this is not likely in the near future in the U.S. The labor movement is in the middle of one of its weakest periods in history, with the smallest portion of workers as union members ever. On the other hand, unions have made great strides in unionizing office and professional workers, though the total numbers are still small.

A more likely motivation for new laws would be evidence of widespread abuse and poor treatment of telecommuters or home workers in general. While there are cases today where people working at home do not receive benefits or rights that they should, I do not believe this is typical. If I am wrong, or if the pattern of abuse spreads, our legislators would be quick to write restrictive laws.

We are in the middle of one of the most interesting evolutionary changes we have seen in the workplace. Telecommuting and the decentralization of work *is* an evolutionary process, in my view, and not a more dramatic revolution. When we consider that it took twenty years for the telephone to achieve a 1% market share in the U.S., we can see how long it takes for traditions to change. When telephones were first introduced, people said they would never become popular - "we have messengers to deliver our information, and besides, people will only want to talk to each other in person." was the reaction. As we know today, we could not live without the telephone - and I think we will say the same about workplace decentralization in years to come.

There are solutions for almost all of the obstacles that can be raised – even the isolation and loneliness said to be experienced by telecommuters, as our friend Homer shows:

DECENTRALIZATION VIA TELETEX
ORGANIZATIONAL AND TECHNICAL IMPACT

EXPERIENCES OF THE RESEARCH PROJECT
"CREATION OF DECENTRALIZED WORK PLACES THROUGH TELETEX"[1]

Barbara KLEIN*
Hans-Peter FRÖSCHLE*

Abstract:

New information and communication technologies allow flexibility with regard to space and time and offer the possibility of developing new organizational forms. The research project "Creation of Decentralized Work Places through Teletex", performed by the Fraunhofer-Institut für Arbeitswirtschaft und Organisation, IAO, investigated different organizational forms like tele-homework, neighbourhood office and exchange of capacities between branches. Main topics were organizational, technical, economic and social impact of telework. The research project provided a unique opportunity of investigating teleworkers who were not freelancers but regular employees. The focus in this essay is on organizational and technical impact, and the views of teleworkers and of in-house employees are presented.

1. Introduction

The situation today is characterized by a dynamic development of information and communication technologies which offer widespread possibilities for decentralization of large areas of white collar work. From an economic point of view there is an additional need to discuss new forms of work organization as part of a trend towards the postulated "post-industrial information society" prior to the general introduction of new technologies. The impact of decentralized forms of work on an individual, operational and social level and their complex interaction are not known. On the other hand there is a lack of scientifically provided empirical research results from which general conclusions concerning the decentralization of white collar work can be drawn.

---

* Barbara Klein and Hans Peter Fröschle are both members of the research staff at the Fraunhofer-Institut für Arbeitswirtschaft und Organisation, IAO, Holzgartenstr. 17, 7000 Stuttgart 1

The expert group "Förderung neuer Kommunikationstechniken (EKOM)" (promotion of new communication technologies) recommended the setting up of a research project "Creation of Decentralized Work Places through Teletex" in order to obtain substantiated findings about the impact of new technologies.

2. CONCEPTION OF THE RESEARCH PROJECT

In the beginning of 1983 the state parliament of Baden-Württemberg commissioned the Fraunhofer-Institut für Arbeitswirtschaft und Organisation, IAO, to carry out the research project "Creation of Decentralized Work Places through Teletex".

The tasks of IAO were to investigate the extend to which

o the lack of qualified typists or office workers in conurbations can be compensated by the establishment of decentralized work places and
o the positive or negative impact to be expected.

The main questions of the study were

o which technical aspects,
o organizational,
o economic and
o social impact

could occur in connection with decentralized places of work.

The investigation was limited in regard to

o the technical system i.e. teletex
o the range of activities (typists) and
o employment contracts (regular employment contracts like employees in the office).

In the planning phase it proved to be difficult to find participants for the research project. A time consuming search was necessary after several companies who had been directly approached by the Ministry of Trade, Technology and Commerce of Baden-Württemberg, decided not to participate in the research project. Their refusal was mainly influenced by a negative attitude on the part of their employees and shop stewards.

In 1983 the IAO asked 65 mainly medium sized companies to join the research project. It turned out that in general there was not only no need to establish decentralized places of work but also the companies did not expect to achieve major advantages or profits. They expected rather to spend a lot of effort and time in adjusting the company´s organizational structure.
Another reason for not participating in the research project was that the establishment of decentralized places of work was too expensive, so smaller firms in particular questioned the profitability.

In the end 14 enterprises and 17 teleworkers participated in the research project.

3. METHODOLOGY OF THE RESEARCH PROJECT

Scientific research concerning the impact of new information and communication technologies in the office area involves many different science disciplines, which in general have different research approaches and focus on different objectives. The necessity of the variety of disciplines is acknowledged, but yet no interdisciplinary theory or concept exists.

A representative survey and random sampling was not possible because no information is available about the diffusion of decentralized work places. The research project was therefore outlined as a pilot study, the methodology was explorative in keeping with the complexity of the investigation. Case studies were performed and qualitative methods were applied.

o The results concerning the social aspects of decentralized work places are based on qualitative interviews and group discussions, which were repeated after a year. They reflect the personal situation of the teleworker. The limited number of cases does not allow any generalization but they are an empirical basis for the formation of hypothesis.

o The results concerning technical, economic and organizational aspects are based on semi-standardized interviews and the transmission of standardized test documents. Common patterns were ascertained which allow a generalization of the results to other applications.

The representation of results systematized by problem fields should not obscure the fact that on the one hand a definite classification by stated problems is not possible. On the other hand complex interaction exists between the areas which also depend on the organizational form of decentralized places of work.

4. ORGANIZATIONAL ASPECTS OF DECENTRALIZED EMPLOYMENT

4.1. Models of Decentralized Work Places

In the middle of the seventies several concepts for locational decentralization of administrative work were developed. Especially notable are the concept of Nilles et. al. /1/ and the congruent concepts of Diebold /2/ and Olson /3/.

Nilles et. al. /4/ identified four evolutionary phases or forms of organizational structures in relation to potential modes of application of telecommunication technologies and their effects on public policy.

o Centralization (current dominant mode in most industries):
Characteristically all administrative operations are concentrated at a single site with workers divided into functional groups according to their primary information product.

o Fragmentation:
In the second state coherent subunits of the central organization break off and relocate elsewhere, though maintaining contact with the parent unit by telecommunications or mail. This phase can actually increase the amount of commuting. The organizational structure of these decentralized units can be either copies of the central organization or in the form of special function units. A starting point could be the search for an optimized combination of production and locational factors.

o Dispersion:
The organization establishes a number of smaller work locations in the area. Employees report to the nearest work centre, irrespective of their special function. Telecommunication and computers connect locations, though executives still requiring face-to-face communication may report to a central place for this purpose. Reduced commuting and improved access to labour markets should result.

o Diffusion:
Premise for the ultimate state of decentralization induced by telecommunication is a developed telecommunications network. In this stage organizations maintain relatively small core staff. Peak work loads may be handled by individual workers who offer their services to several clients or organizations through telecommunication networks.

The categories of Olson /5/ and Diebold /6/ are characterized by a mononuclear view. They categorize several feasible alternative work arrangements which inject a degree of flexibility in the locational and temporal definitions of work:

o Satellite work centres:
At this stage a relatively self-contained organizational division is relocated physically. In contrast to "fragmentation" the focus is on optimizing commuting times for all employees.

o Neighbourhood work centres:
Employees from different organizations share space and equipment in a work centre close to their homes.

o Flexible work arrangements:
This concept tries to meet different interests of professional and family demands. Job arrangements can be flex-time and job sharing.

o Work at home:
This concept is an extreme case of individual work options. Here employees work on a regular basis at home. This concept corresponds to Nilles levels of "dispersion" and "diffusion".

Schäfer /7/ states that by joining these category systems a negative correlation between the degree of locational decentralization and the degree of functional absorption can be stated.

4.2. Decentralization performed in the Research Project

In the research project following forms were implemented:

o   work at home

o   a neighbourhood office
    Two women were sharing a place of work, which was set up in the house of one of the employees. They were writing for three institutes, but belonged to the hierarchy of the administration, which was placed at a different location.

o   capacity exchange between branches
    The structure of the organization was modified in so far that teleworkers belonged to the hierarchy where their workplace was located. But orders from professional workers of the other branch had first priority.

These practical forms of decentralized work had the focus on location i.e. with reduced or no commuting time.

Although each of the investigated offices had its own characteristics, they had certain organizational features in common. Before decentralized places of work were installed the organizational pattern of the typing pool can be described by Figure 1:

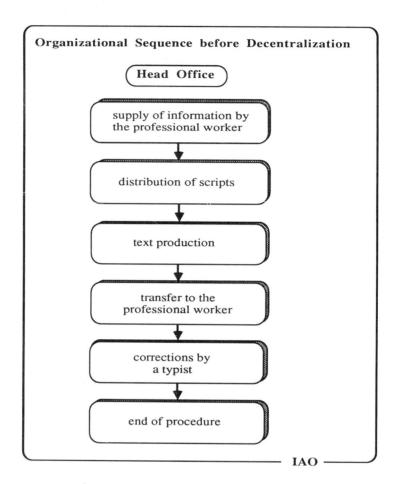

Figure 1:
Organizational Sequence before Decentralization

The professional workers provided scripts in form of manuscripts, cassettes and files to the typing pool. The head of the typing pool distributed them among the typists. (This distribution was handled quite differently in each typing pool, e.g. the head of the typing pool distributed the work randomly to any typist with unused capacity. In one of the typing pools the division of labour was very high, so certain typists produced continuous texts, others just modular texts, others were responsible for texts requiring an intimate familiarity with the system.)

After the typist had produced the text, scripts and text were transferred to the professional worker. S/he revised the texts and transferred them to the typing pool where they were corrected as a rule by the typist who had produced them.

Figure 2 shows the main features of the organizational sequence after establishing decentralized places of work.

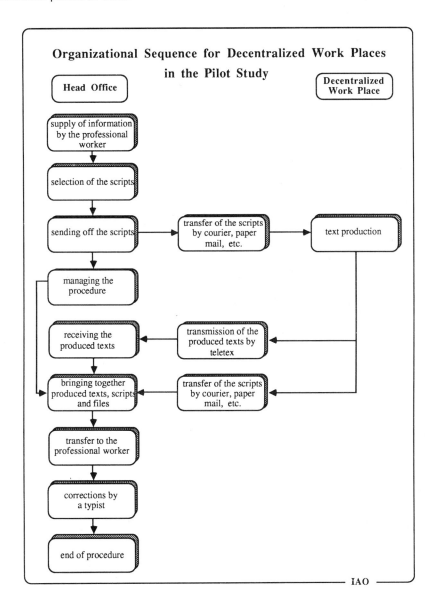

Figure 2:
Organizational Sequence for Decentralized Work Places in the Pilot Study

After the supply of scripts in form of manuscripts, cassettes and files, the head of the typing pool had to select which of the scripts were suitable for the decentralized work place.
Criteria were

o the expected size of the document:
   as a rule large documents were more suitable,

o the degree of difficulty:
   the text should not be specific so enquiries could be reduced to a minimum,

o urgency:
   the texts should not be urgent since transferring the scripts by courier or paper mail entailed delays of up to three days,

o data protection:
   criteria of data protection had to be considered; e.g. sensitive or confidential items could not be sent in a regular manner.

After having selected the scripts, they were transferred to the decentralized work place either by courier, paper mail or pick up. The teleworker produced the text according to the scripts. Then the text was transmitted via teletex to the head office, whereas the original scripts were returned the same way they came.

In the head office the teletex operator received the text, stored it and made a hard copy.

Then the text was either transferred to the professional worker or, in case of circulating files, the text was stored until the scripts arrived. Files often had to be circulated because typists had to look up spelling of names or copy parts of the files. Then the head of the typing pool had to bring the produced text, scripts and files together and transfer them back to the professional worker who usually needed both, text and files, for further processing.

After s/he had revised and transferred the text back to the typing pool, the corrections were assigned to a typist of the typing pool.

4.2.1. Viewpoint of the Head of the Typing Pool

All heads of the typing pools mentioned the additional work due to the decentralized work place, but their personal views differed. Some just mentioned the additional work, others stated a gain in flexibility to work on urgent texts or other duties in the typing pool. By using the capacity of decentralized work places for the production of large documents, in some cases it was possible to relieve the typing pool, so they could focus on the very urgent work.

Although all forms of decentralization had the same pattern, the extent of additional effort varied.
The managing effort for work at home and the neighbourhood office was far more intensive than for the capacity exchange between branches.

In accordance with the employment contract, telehomeworkers were not at the enterprises disposal but had to work a certain amount of time each day like the typists in the office. For the head of the typing pool in the head office this implied occupation of the teleworker on a regular basis.

It was difficult to achieve a normal degree of capacity utilization. The selection of scripts in particular caused severe problems. As a rule the scripts were either not long enough or urgent, so it happened quite often that there was not enough decentralized work to occupy teleworkers at home or in the neighbourhood offices within the agreed hours.

Capacity exchange between branches was much easier to deal with. First the head of the typing pool in the head office did not have the pressure to send a sufficient amount of scripts just to occupy teleworkers. Instead the teleworker - although working in priority for the head office - worked for the home branch if there was nothing to do for the head office. A positive effect was that their work was not limited to text production for the home branch because they not only did text processing but clerical work, too.

4.2.2. Viewpoint of the Teletex Operator

From the point of view of the teletex operator in the head office the enforced division of labour was judged negatively. Corrections were done in the head office to save time. So typists not only had to do their own corrections but the corrections of the teleworker as well.

To do corrections of the teleworker the typists needed a certain amount of time to acclimatize. In addition great effort was necessary to give the text a logical structure, which cannot be transmitted by teletex if no private use mode is installed.

The fact of being occupied with correction work much more than before decentralization was judged as a loss of self-esteem. In their opinion a good typist is one who has to do very little correction work. However, now it was possible that they had to correct for days on end.

The typists did not share the positive view of the head of the typing pool regarding the relief of the typing pool by decentralized places of work. Moreover they criticized the additional strain especially because of the involvement of so many people in processing a text.

4.2.3. Viewpoint of the Teleworkers

Teleworkers criticized that their work was restricted to text production. Because they could not do any corrections they felt that they might loose qualification. Usually they did not get any feedback from either the head of the typing pool or professional workers.

Although the employment contract of the telehomeworkers guaranteed that they had to be at disposal within the agreed hours, they usually met internal requirements, especially because of a feeling of obligation due to previous inactive periods. In case of a deadline they would work overtime as long as they could combine it with family requirements.

Other projects which investigate telework also mention the problem of the degree of capacity utilization, albeit from a different point of view. The study of Goldmann and Richter /8/ shows that less than half of the investigated women got work on a regular basis. Huws /9/ also stated that about a third of the telehomeworkers in her sample often had periods without work.

In the research project teleworkers received their wages even if their degree of capacity utilization was less than their working hours because their status was identical to the staff in the office. In contrast the characteristic feature of the situation of most teleworkers in above stated projects is that they have no employment status and the risk of having work or not is transferred to the teleworker, whereas on the conditions of the research project the risk rested solely on the employer.
So managers criticized a lack of profitability but were unable to take any action to improve the efficiency of decentralized work places.

5. TECHNICAL ASPECTS OF DECENTRALIZED WORK

The research project was based on the assumption that teletex is more than just a substitute for telex. The main issue was whether teletex is qualified for decentralizing clerical work.

5.1. Characteristics of Teletex

In 1982 the Federal German Post Office introduced the teletex service officially. The recommendations of the Comite Consultatif International Telegraphique et Telephonique (CCITT) are the basis for the international standard. The service allows the sending of documents which have been written on a teletex machine to teletex and telex machines.

Figure 3 shows the structure of a teletex terminal and the characteristics of the local and the communication part /10/.

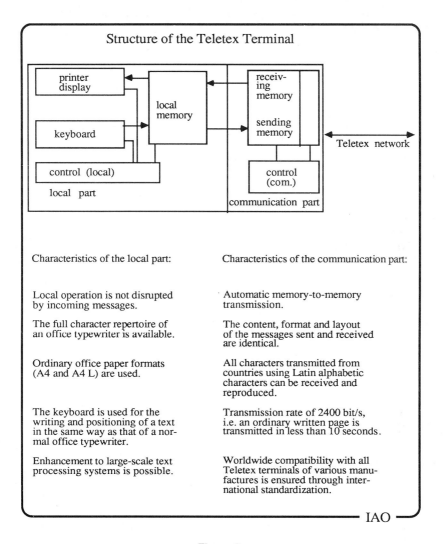

Figure 3:
Structure of the teletex terminal /11/

Requirement to participate in the teletex service is device compatibility. Received documents have to correspond in content as well as in layout to the transmitted document.

A transmission which corresponds in layout to the initial document does not include the transmission of the logical structure of the initial document. This implies that control functions e.g. carriage returns, tabulators, are usually not transmitted. These control functions are of relevance if corrections are necessary. If they lack, page formatting implies a lot of difficulties.

The Federal Post Office was asked about the extend to which control functions have to be transmitted. In order to obtain a licence for teletex, it is sufficient to prove that transmitted documents look like the original text when printed.

However, some suppliers offer the private use mode as an additional application. This private use mode allows the transmission of control functions between identical systems with identical software versions. Applying the private use mode the logical structure of the document remains after transmission.

5.2. Experiences with the Teletex Service

Interviews were done in regard to the application of teletex with

o teleworkers,
o teletex operators in the head office,
o managers.

In addition standardized documents had been transmitted between identical and different systems in order to test the transmission function.

Many technical problems arose especially in the introductory stages. In some cases the standard was not fulfilled and a transmission corresponding to the layout standard was not possible. In some cases the systems had to be exchanged several times before regular work could start.

5.2.1. Experiences of the Teleworkers

The teleworkers (regardless of the kind of decentralized places of work) used the service nearly exclusively in its transmitting function.

The following technical restrictions occurred:

o In some cases the memory for receiving and transmitting documents was too small. This meant that the text to be transmitted had to be divided into separate smaller documents. For each of these smaller documents a new transmission sequence had to be started which implied higher transmission costs.
In a few cases it was necessary to coordinate times when documents were to be transmitted between the decentralized work place and enterprise so the memory of the system in the enterprise could be cleared and work interruptions minimized.

o Problems while transmitting documents:
Some documents got lost although the reference information (which provides identification of called as well as calling terminal, date, time and supplementary reference information) kept a record of successful transmission.
With the memory typewriter Eritex 10 documents had to be written in a special so-called "teletex-mode". This mode does not indicate e.g. whether a page has more characters than it is supposed to have. If this was the case the transmitting process was interrupted, but the system did not show the error.
To transmit the document it first had to be reformatted.

Some of the systems did not have the option of automatically delayed transfer which was desirable to cut down fees by transmitting at off-peak rates. As a consequence a few telehomeworkers transmitted the documents late in the evening out of regular office hours.

5.2.2. Experiences of Teletex-Equipment Operators in the Head Office

The teletex-equipment operators in the enterprise used both the transmitting and the receiving function of the service.

Following technical problems were mentioned:

o Compatibility of systems:
The effort of revising transmitted documents depended on the compatibility of the systems.
If the system had no private use mode, the staff had to revise the document to give it a logical structure. Two examples are given to demonstrate the effort to give a document its logical structure:

Carriage returns had to be filled in at the end of paragraphs on the Philips P5020. In addition the special hyphen-function did not work. Divided syllables were not recognized as belonging together if characters were inserted or deleted. The typist had to look for dashes through the whole document and correct it.

The revision of documents transmitted by any other system besides an IBM 6580 to an IBM 6580 entailed with many difficulties. On incoming documents line spacing of 1.5 was depicted with two lines on the terminal. The printout corresponded to the input of the system which sent the document. If characters were inserted in the transmitted document the text shifted in so that the inserted characters had a different line spacing than the rest of the text. Revising the document implied that line spacing had to be modified.

o Standards of the service:
The Federal Post Office prescribes that transmitted documents have to be "visualized" before starting revision. In order to fulfill this requirement documents received on a memory typewriter had to be printed before any revision could start. This was particularly annoying because the reference information is usually on top of each page. This reference information cannot be deleted until after the printout.

The standard should provide for undisturbed local mode operation as well as an undisturbed communication function. Although it is possible to continue working with the system while receiving a document (some of the systems slowed down considerably), in most cases interruption of work was necessary. The visual/audible signal "message received" gives neither information as to whether a telex or a teletex has been received nor whether the message is urgent or the size of the incoming document. Clearing of memory is of major importance, especially for maintaining a permanent receiving memory.

5.2.3. View of the Managers about Teletex

The managers mentioned following criticisms:

o Cost benefit relations
The structure of the teletex charges is designed for a large amount of communication. Monthly basic rentals are high, but additional transmission charges are low, whereas the charge structure for the telex service consists of a low basic rental and high transmission charges. The charge structure does not correspond to the actual diffusion of teletex. There are 15, 000 subscribers to the teletex service in the Federal Republic of Germany and only 13 other countries offer the teletex service. The telex service is offered worldwide with a 160, 000 subscribers in the Federal Republic of Germany. Although it is possible to send a telex by a teletex system (network interworking is performed via a conversion facility in the network modes which regulates codes, protocolls and transmission speed) costs rise because when transmitting a telex the high transmission charges for telex have to be paid as well.

A negative cost benefit relation results also from the fact that to overcome incompatibilities users had to install identical systems and identical software versions with private use mode. Here a reduction in costs through installation of cheaper systems at the decentralized work place could not be effected. If users had installed different systems they had to accept that typists had to spend a lot more time on correction work.

One of the users complained that transmission speed of 2400 b/s (i.e. approx. 10 sec. per page) is too slow, so costs are too high for transmitting larger amounts of documents.

o Consultancy
A few users criticized a lack of consultancy of the Federal Post Office and suppliers. If technical problems occurred, down-times were too long, no solutions of problems were developed.

o Service standard
One user criticized the lack of a multiaddress service as it is offered by the telex service with defined subscribers. Multiaddress service means a radial transmission of a telex to several communication partners by a single transmission order.
Although some of the teletex systems offer the option "multiaddress service" this means a serial transmission of transmission orders to several communication partners.
In addition it is not possible to receive a telex by multiaddress service with a teletex system.

o Compatibility
   Users hoped to overcome diskette incompatibilities with teletex. But very soon it became obvious that the basic requirements of the teletex service were not sufficient for the application field if documents had to be revised after receipt. The private use mode certainly has this option but identical systems and identical software versions then have to be installed.

Managers thought positively about the functionality of the systems which was not only restricted to the communication function, but could also be used as a word processing system.

As a conclusion it can be said that teletex can be used as a technical medium to decentralize typists' work, but the extend to which the additional effort of doing corrections (estimations were 20-30% more time spent than on usual corrections) is justified is a matter for debate.

There is also the question whether the gain in flexibility by quick transmission in private use mode between identical systems justifies the costs for a decentralized connection or whether an exchange of diskettes will meet needed requirements.

6. ECONOMIC ASPECTS

At present there are no instruments for an economic assessment of decentralized organization forms which consider classical cost accounting as well as non-monetary criteria (i.e. increased flexibility, maintaining the qualification of workers).
A comparison of cost accounting models between alternative forms of organizations showed, on a level of isolated profitability, that costs per page (on the basis of identical preconditions: typing performance, using the capacity of the systems, personnel costs) were higher for decentralized work places than for places in a typing pool. Factors which can differ are costs for the communication system and communication fees. Personnel costs were not relevant as a differentiating factor because of the identical employment status of all staff members, no matter if employed centrally or decentrally.

Under these circumstances the capacity utilization rate of decentralized work places was the critical factor for the cost development. The places of telehomework were the most expensive alternative compared to other decentralized places of work, especially for part time work. In the neighbourhood office and in branches the degree of capacity utilization was higher because of locational concentration of employees where systems could be used by colleagues during holidays or illness.

Thus decentralization of workplaces was involved with too many problems which resulted in a negative assessment of costs and benefits.

Besides these general statements about profitability, no differentiated results were achieved. Because of consideration of data protection a detailed investigation about processing time and performance of typists could not be carried out.

As far as the teleworkers were concerned the research project showed that, in accordance with their employment contracts, they had no financial disadvantages compared to their in-house colleagues.
The study of Huws /12/ shows that 67 per cent of teleworkers in her sample felt that they were earning less than if they had been doing the same work in the office.
Major economic disadvantages are that the responsibility of having work or not is shifted to the teleworker when they are paid by piece-rate or when they are freelancers. Teleworkers without employment contracts also usually have no protection in the field of social security either and do not receive unemployment benefit in periods without work. Also they quite often have to buy their own technical support system.

7. SOCIAL ASPECTS

In the research project all workers at decentralized work places were women. Social impact involved with these jobs included problems which are generally associated with working women with family commitments. Major problems arose due to stress within job, family and housekeeping as well as potential role conflicts between family members.

The effects of traditional homework were also observed. Important was the lack of a separation between job, family and leisure time as well as problems of integration of telehomeworkers into the head office.

Because of the small numbers of participants in the research project only limited generalizations are possible.
In general it may be said that the potential negative impact for each individual and society seems to be less within decentralized forms such as neighbourhood offices and capacity exchange between branches.

o In most cases telehomework cannot be regarded as an instrument to overcome traditional gender role allocation, which assigns women primarily to child care responsibilities and reproductive functions. The women usually had additional stress because of job, housekeeping and children and were in a conflict to master all these fields. Because she was always present at home a renewed distribution of family work did not occur.

o The decision to do telework was judged as a decision for the family as well as for the job. Working hours and locational flexibility allowed an "optimal" combination of family work and professional activity. These organizational forms of working were an alternative in a transitional phase of intensified family work. Nearly all teleworkers would not have done this job if they had had no children. During the stage of primary child-care responsibility the family area is of greater importance than a professional career.

Other research projects /13/ also mention the combination of employment and child care responsibilities as a main motive to do telehomework .

Another criterion for starting this kind of work was the fear of losing professional contact and contact with technical developments.
Workers who lived in rural areas had either no alternative to find equivalent work or had to take long commuting times into account.

o In the initial period integration into the head office only involved instrumental communication about work flow and technical facts. This rare communication was reduced even more in course of time. A communication barrier was built up which made it difficult to communicate especially if teleworkers were specially employed for the duration of the research project.

o An own room to work did not help to support a separation of family and professional work. Family demands, especially children of nursery school age, allowed undisturbed professional work only if working hours were handled very flexibly. The working hours were divided into multiple parts. The daily routine was planned accoording to demands of the employer and those of the family. Telehomeworkers usually attempted to keep a fixed time for the first part of professional work and adapted the rest of the work according to family demands.

o All teleworkers judged the flexible working hours positively because it allowed flexible reactions in emergency situations. But in some cases (especially if telehomeworkers had infants) undisturbed professional working was only possible very early in the morning or late in the evening.

This "flexibility" turned out in fact to be fictitious flexibility and the daily routine became concentrated because of the permanent change in professional and family work.
An employment contract guaranteed that the teleworkers had to be available only a certain amount of time per day according to their contract. But other forms of payment (payment by piece-rate, freelancers) focus on the company's flexibility.

The arrangement of employee working times in branches and the neighbourhood office depended on organizational requirements which could not be influenced. In some cases there was the possibility of arranging working hours in agreement with colleagues.

The social impact of decentralized employment did not result from the implementation of information and communication technology. The observed problems were comparable to those met in traditional homework in the field of text production /14/.

The general positive attitude of teleworkers to their work, confirmed by research projects /15/, could be interpreted as a pragmatic adaptation to the situation.

It lies within the character of their situation that the alternative to telework is usually not a qualified secure employment in the enterprise but rather unemployment.

8. FUTURE PROSPECTS

There are several enthusiastic predictions for telework. In 1971 American Telegraph & Telephone (AT & T) predicted that all Americans would be working at home by 1990. In 1980 Toffler /16/ predicted half of the white collar worker could work at home.
In the beginning of the eighties the diffusion of telework was negligible. In research projects of the European Foundation /17/ and the Battelle-Institut /18/ no teleworker could be found within the Federal Republic of Germany. Meanwhile several projects and pilot studies have been implemented. An evaluation of these studies indicated about 300 teleworkers /19/.

How will the diffusion of telework look like in the future?

Given the technical possibilities there are still technical constraints and restrictions which have a negative impact on economic and organizational factors concerning decentralization on a wider scale.

At present no multifunctional data systems are available which allow access by one data terminal to different communication services of the Federal Post Office. Decentralization of managerial, professional, clerical and secretarial work requires communication, processing and storage of language, data, text and graphics and moving images. Today available services allow the transmission of language, data, text, freezing frames and graphics, but transmission is accomplished by separate channels and separate data terminals.

The narrow band ISDN (integrated services digital network), which is a further development of IDN, will offer a single line integration of the main electronic services which are offered by the Federal Post Office.
For most communication forms one interface will then suffice. Provided that the policy of the Federal Post Office remains constant, high access fees and costs of operation are not expected.

The development of multifunctional ISDN-data terminals allows the transmission of multimedia documents, which will certainly influence the in-plant as well as the intercompany area. Hitherto no organizational concepts of decentralization have been available which describe locational as well as organizational decentralization (as changed distribution of competences and along with this the establishment of a working process which is to a high degree autonomous, whereby autonomy should not be understood as segmented work parts but as integrated working of a process) for different areas.

The decentralization of typist work according to the European Foundation /20/ can be regarded as an intermediate stage in a process of complete automation where typists will eventually be replaced by a computer having voice input capability.

In terms of numbers, diffusion of telework will be a low-scale affair. Problems of telehomework should be regarded in connection with future problems involving redundancies due to technological advances.

9. CONCLUSIONS

The research project "Creation of Decentralized Work Places through Teletex" showed that in principle it is possible to establish decentralized work places and that, given the technical systems existing today, a shortage of qualified typists in conurbations can be compensated by teleworkers.

The main results of the investigated aspects were:

o   The Teletex service and data terminals can be judged as reliable Problems arise when transmitted documents are corrected.

o   Major organizational problems are caused by the fact that transfer of scripts is still restricted to traditional ways (courier, paper mail) because electronic alternatives are still too expensive today and have limited capabilities.

o   Costs of decentralized work places are higher than the traditional work place of a typist. Factors influencing profitability are technical support systems, transmission fees and the rate of capacity utilization.

o   Social impact concerned gender role allocation due to additional stress caused by the need to master job, childcare responsibilities and housekeeping.

The positive assessment of teleworkers concerning their decentralized work places indicates a need for both flexible work places and flexible working hours. Problems arising were solved on an individual basis.

In contrast to the positive assessment by the teleworkers themselves, managers and staff in the head offices directly concerned with decentralized work had a rather negative attitude towards telework as a result of technical and organizational restrictions which implied additional stress and effort for them.

The results of the research project indicate that the diffusion of decentralized work places in this domain will be negligible. Decentralization of typist work will be limited to definite problems and individual instances.

In general a diffussion of decentralized work places will be determined by the availability of appropriate technical support systems and by the relation between demand for profitability on the part of the enterprise and social security requirements on the part of the teleworker.

NOTES

1   This article is based on an unpublished detailed report of the reseach project "Creation of Dezentralized Work Places through Teletex":

Fröschle, Hans-Peter, Klein, Barbara, Schaffung dezentraler Arbeitsplätze unter Einsatz von Teletex, Abschlußbericht, (July 1986)

It can be ordered from the Fraunhofer-Institut für Arbeitswirtschaft und Organisation, IAO, Holzgartenstr. 17, 7000 Stuttgart 1

REFERENCES

/1/   Nilles, Jack M., Charlson, Roy F., Gray, Paul, Hanneman, Gerhard J., The Telecommunications-Transportation Tradeoff, (New York 1981)

/2/   Diebold Group Inc., Office Work in the Home: Scenarios and Prospects for the eighties, "The Diebold Automated Office Program", (New York 1981)

/3/ Olson, Margarethe H., Remote Office Work, Implications for Individuals and Organization, (New York 1981)

/4/ cf. Nilles et al., op. cit., (1981) pp.11-17

/5/ cf. Olson, op. cit., (1981) pp. 8-12

/6/ cf. Diebold Group Inc., op. cit., (1981)

/7/ Schäfer, Peter, Technisch-organisatorische Aspekte der räumlichen Dezentralisierung von administrativen Tätigkeiten, graduate study, (1986) p. 24

/8/ Goldmann, Monika; Richter, Gudrun, Teleheimarbeit in typischen Frauenarbeitsbereichen - Arbeitspapiere III: Teleheimarbeiterinnnen in der Satzerstellung/ Textfassung für die Druckindustrie, (Sozialforschungsstelle Dortmund 1985) unpublished manuscript, pp. 63-69

/9/ Huws Ursula, The New Homeworkers, (London 1984) p. 33

/10/ Schenke, Klaus, Rüggeberg, Rolf, Otto, Jens, Teletex, a new international telecommunication service for text communication, (Bad Windsheim 1981)

/11/ Schenke et al., op. cit., (1981) p. 11

/12/ cf. Huws, op. cit., (1984) p. 36

/13/ cf. Goldmann et al., op. cit., (1985) p. 12 ff.
cf. Huws, op. cit., (1984) pp. 28-29
cf. Olson, op. cit., (1981) p. 27 ff.

/14/ cf. Haas, Hans-Dieter, Scherm, Georg, Die Rolle der Heimarbeit auf der Schwäbischen Alb, Kullen, Siegfried (ed.), Aspekte landeskundlicher Forschung, Festschrift zum 60. Geburtstag von Hermann Grees, (Tübingen 1985)

/15/ cf. Goldmann et al., op. cit., (1985) p. 15 ff.,
cf. Huber, Joseph, Telearbeit - Eine futuristische Fiktion als Politik, (1985) unpublished manuscript

/16/ Ballerstedt, Eike, Telearbeit, Die neue Gesellschaft, Frankfurter Hefte, Nr. 3, (1985) p. 219

/17/ European Foundation for the Improvement of Living and Working Conditions, Telework, Impact on Living and Working Conditions, (Dublin 1984) p. 25

/18/ cf. Ballerstedt, Eike et al., Informationstechnisch gestützte Heimarbeit - Studie über Auswahl, Eignung und Auswirkungen, (Frankfurt/ Main, Tübingen: Battelle/ Integrata 1982)

/19/ cf. Schäfer, op. cit., (1986) p. 29

/20/ European Foundation, op. cit., (1984) p. 41

4

FUTURE PERSPECTIVES AND STRATEGIES

TELEWORK – POTENTIAL, INCEPTION, OPERATION AND LIKELY FUTURE SITUATION

Werner B. Korte

empirica GmbH
Kaiserstr. 29-31
D-5300 Bonn 1

1. INTRODUCTION

Advances in the development and application of information technology will lead to major changes in the European and world labour market in the coming decades. Basically the dispersal of office work over time and space is feasible. In the future, the more office functions become computer based and the more computers become linked by communication networks, the less it will be necessary for office work to be carried out in the same office or even the same building. Innovations in information and communication technologies have brought forth a novel form of work organization which has received significant attention: telework.

The paper starts with a brief overview of forecasts of the extent of telework followed by a characterization of what telework entails and the various current forms of this new work organization. Section 4 deals with an analysis of awareness and interest in telework in the major European countries based on surveys of the workforce and company decision makers. Section 5 presents the results of case studies of European companies practising telework. The focus will be on reasons and motivation of company decision makers and individuals for starting and taking up telework and factors facilitating and constraining the uptake of telework from the point of view of both groups. It will be followed by two classifications of telework inception and schemes as a result of different actors and strategies for implementing telework. The paper ends with the formulation of tentative conclusions.

Figure 1 presents an overview of the logical steps of the present paper.

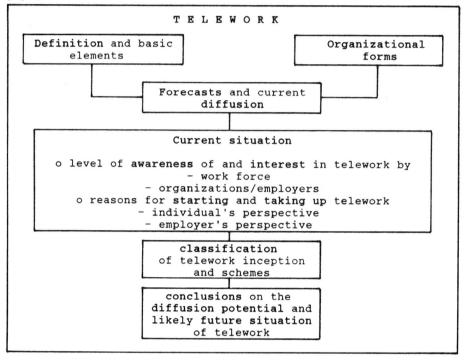

FIGURE 1:
Logical Steps of the Present Paper

2. FORECASTS OF THE DIFFUSION OF TELEWORK

Technology offers the opportunity for significant changes in both job content and the concept of a (central) office. It is against this background that a number of forecasts especially of the diffusion of electronic homework have been undertaken in the 1970s and early 1980s.

In the early 70s (1971), AT&T predicted that in 1990, all Americans would be working from home [12]. Alvin Toffler's (1980) forecast still reaches 50%, whereas the Institute of Future Studies at the University of South California assumes a number of 20% in 1990 and 40% in the year 2000 [16]. The numerous predictions of the past about the potential development and spread of telework have one thing in common: they all overestimated the speed of diffusion of this new work organization based upon IT. The wide divergence between these estimates and the actual extent reflects the considerable confusion about the meaning of "telework". We can only give a very rough overview of earlier development and the present position of telework. These are illustrated in the following figures which indicate a rather small current penetration rate of this new work organization.

```
o  Federal Republic of Germany         > 1,000 teleworkers
o  France                              < 1,000 teleworkers
o  United Kingdom              approx.   3,000 teleworkers
o  Italy                               < 1,000 teleworkers
o  U.S.A.                      approx.  10,000 teleworkers
```

FIGURE 2:
Telework Penetration Rate (estimates) 1987

```
1982: Telehomework projects are as hard to find as pins in a
      haystack (Battelle/Integrata, 1982) [2]

1985: 50 - 100 telehomeworkers (Ballerstedt, 1985) [1]

1986: 1000 teleprogrammers (not including teleworkers perform-
      ing other tasks) (Heilmann, 1986) [10]

1987: at least one thousand telehomeworkers (empirica, 1986) [3]
```

FIGURE 3:
Diffusion of Electronic Homework in the Federal Republic of Germany

In summary, "it is safe to say that telework has not lived up to its expectations" [9] but that it shows a gradual, evolutionary growth.

3. DEFINITION AND ORGANIZATIONAL FORMS OF TELEWORK

In order to put the discussions about telework on a common footing, it seems essential to have a clear (working) definition of the term, or at least a common understanding of the basic elements characterizing telework.

3.1 Definition and Dimensions of Telework

It does not seem appropriate to define telework along any single parameter or dimension. Form of work organization, place of work, contractual arrangements, etc. are all relevant or necessary parameters in a definition of telework, but are not sufficient if seen in isolation.

Telework has to be seen in the frame of some general changes currently taking place in organizations. In response to growing competition and market pressures there is an increasing trend towards decentralization on a number of levels, allowing for the higher flexibility demanded by the market. Decentralization is occuring at the levels of

- _establishments_, resulting in a shift of competence and responsibility to middle management and employees;
- _enterprises_, characterized by the formation and establishment of smaller subunits, e.g., branch offices and

- **economies**, taking the form of externalization of labour by larger companies to smaller services companies or individuals working from their homes.

Telework involves three basic elements:

1. **Location**
   The location of the work site is determined by the needs of the teleworkers and/or organizations and is relocatable as desired or needed. This implies that the geographical site at which work is completed is independent of the location of the employer and/or contractor.

2. **Use of IT**
   Telework relies primarily or to a large extent on the use of IT (PC, storage typewriter etc.).

3. **Communication Link to Employer/Contractor**
   3.1 Telework - Narrow Conception
   A communication link exists between/among the teleworker and the employer/contractor which is used for electronic communication and transmission of work results.
   3.2 Telework - Broad Conception
   The teleworker works at a distance (spatially separate) from his/her employer and/or contractor whereby work results are stored on a disc, cassette etc. There is no electronic communication link used for data transmission. The work results are delivered by traditional media, such as mail, courier etc.

3.2 Organizational Forms of Telework

Today, a number of different organizational forms of telework can be observed [18]. Electronic homework is an extreme case and the most decentralized form. Those forms commonly described and discussed in the current literature are as follows:

1. Satellite Work Centres

   Relatively self-contained organizational divisions in a firm which have been physically relocated and separated from the parent firm. The emphasis is on locating these centres within a convenient commuting distance for the greatest number of employees utilizing the site. The supervision of work is generally by management staff on site.

2. Neighbourhood Work Centres

   Offices equipped and financially supported by several companies or organizations. In these offices, employees of the founding organizations share space and equipment in a location close to their homes. Supervision of work is carried out remotely.

3. Flexible Work Arrangements

   Provide employees with flexibility in the scheduling and location of work. This option recognizes the need for occasional alternative work arrangements, especially for professional and managerial employees and provides mechanisms to accommodate family as well as work responsibilities.

4. Electronic Homework

The most decentralized form of telework. Employees work at home on a regular basis. While homework depends virtually completely on remote supervision and does not provide a field for work related social interaction, it does offer employees maximum flexibility in scheduling working time.

5. Electronic Services Offices

Interdependent firms which carry out a wide range of data processing and computer related services for small and medium sized firms. Larger companies also make use of such services in times of internal bottlenecks or peaks.

4. TELEWORK POTENTIAL - AWARENESS OF AND INTEREST IN TELEWORK IN THE MAJOR EUROPEAN COUNTRIES

Given the currently very low penetration rate of telework throughout Europe the question arises of whether this new work organization is likely to remain marginal. We have therefore tried to shed some light on some essential aspects affecting the uptake of telework: the awareness of and interest in utilizing telework by the work force and decision makers in companies as well as the attitudes of the social partners towards telework. The present paper restricts itself to a presentation of the former.

The empirical data which forms the basis of the analysis was obtained in two representative surveys:
o the Employed People Survey (subsequently referred to as EPS) inquired into the interest of 16,000 employees in electronic homework [4] and
o the Decision Maker Survey, which revealed information on the interest of 4000 decision makers in companies in telework [5].

4.1 Awareness of and Interest in Telework by Employees

According to the results of the Employed People Survey, approximately 13 million employees in the four major European countries are interested in electronic homework [4,14]. These figures have to be compared with the overall work force of 92 million of which only the work of approximately 50 million is theoretically suitable for being performed in telework arrangements. Thus, about 25% of the employees whose work is suitable for being performed in such a work arrangement are interested in electronic homework. This constitutes a very significant awareness of and potential for telework. In addition to recognition of the suitability of tasks for telework, acquaintance with new information and communication technologies plays a decisive role in determining attitudes towards telework. Accordingly, interest is significantly higher among workstation users, home-computer users as well as among employees working in occupations which are the most highly penetrated by new information and communication technologies. As illustrated in Figure 4, interest in electronic homework of workstation users ranges from almost 1/5 of the work force in the Federal Republic of Germany to 1/3 in the United Kingdom. The figures are even slightly higher among the home-computer users and substantially higher among DP-professionals already using a workstation, ranging from 35% to more than 60% in the United Kingdom.

	Federal Republic of Germany	France	United Kingdom	Italy
Employees total	9%	14%	23%	11%
Employees with jobs suitable for telework	13%	27%	32%	28%
Workstation users	17%	28%	33%	21%
Private PC-users	17%	33%	35%	31%
overall level of interest	low/medium	medium	medium/high	medium

Source: empirica [4]

FIGURE 4:
Interest of Employees in Telework and Acquaintance with IT (in %)
(Percentage of all employees in a category professing interest in engaging in electronic homework)

Based on the Employed People Survey results, one can draw a preliminary profile of the European employee most likely to be interested in telework.

Age	1. 20 - 29 years 2. 15 - 19 years
Occupational Status	1. self-employed 2. full-time employee
Household Structure (no. of earners)	1. 2 earners 2. 1 or 3 earners
Work Content and Profession	1. DP-professionals 2. business and other professionals 3. scientific/engineering + secretaries, stenographers, typists
Company Size	1. 100-249 employees 2. 1 employees 3. $\geq$1000 employees

Source: empirica [4,7]

FIGURE 5:
Strata of the European Work Force Interested in Telework

Accordingly, he/she
o is young (15 – 30 years of age),
o is self-employed or a full-time employee,
o lives in a household with at least two working people,
o works in an occupation already exposed to computer work and requiring high qualifications,
o either works as an individual entrepreneur or in a large company,
o works in no particular industry.

4.2   Awareness of and Interest in Telework by Decision Makers in Companies

The Decision Maker Survey (DMS) found that, apart from those in Italy, the majority had reserved attitudes towards telework. Throughout Europe, the decision makers in companies generally show little awareness of the potential advantages of telework applications. However, the proportion of decision makers showing interest in telework for specific, suitable tasks is significantly greater than the proportion actually implementing telework. For these, there is clearly a relatively low initial barrier to take up [4,20].

	Proportion of decision makers		showing no interest regardless to tasks
	showing interest in telework for least suitable tasks	most suitable tasks	
Federal Republic of Germany	1%	18%	56%
France	9%	17%	55%
United Kingdom	2%	3%	65%
Italy	8%	43%	34%

Source: empirica [5]

FIGURE 6:
Demand for Telework by Decision Makers in Companies (in %)

The tasks identified for widespread use of telework include: typing and word processing, data entry and amendment and computer programming, as well as clerical work in particular branches of industry. All these tasks are forms of office work which meet particular job characteristics (e.g. low need for communication, defined outputs) which are suitable for performance of a job in telework arrangements [11,17,18].

4.3   Synthesis

Combining the results of the two mass surveys, there seems to be a general awareness of and interest in telework and accordingly a potential for decentralization of certain activities. Without extrapolating from subjective acceptance to practicability, it can at least be said that roughly 25% of those whose work could be decentralized would be happy with such a development. Major variations occur with regard to the level of education of employees and between industries. Above average interest in telework is being shown by EDP specialists, engineers, secretaries/ typists, clerical workers, accountants, lawyers and journalists, as well as commercial sales persons. This latter category has, however, a long tradition of decentralized work as "travelling salesmen".

Specific industries which stand out as having the greatest potential need and utilization rate are: banking and insurance, the wholesale and retail trade, the private service sector as well as freelance professionals.

Small and medium-sized establishments (SMEs) have a need for external specialist services such as financial planning, payrolling, accounting, etc. Computer programming and other specific EDP activities are also of interest to SMEs in particular, whereas larger companies tend to keep these functions in-house.

To sum up, and as illustrated in the following figure, it can be said that with regard to work content, tasks and jobs in the European economies in which IT is already used to a considerable extent are likely to be performed to a growing extent in telework arrangements in the future. Moreover, industries best equipped and acquainted with IT today as well as large organizations, but also the self-employed (e.g. entrepreneurs, freelance professionals) are likely to become the frontrunners with this new work organization based upon IT.

	Supply of Teleworkers	Demand for Teleworking
Work Content/ Occupational Group	o DP-professionals o business professionals o scientific/engineering o secretaries, stenographers, typists	o computer programming o typing, wordprocessing o clerical and administrative work o data input/amendment
Industry	o finance o process	o finance o private services o distribution
Company Size	100 - 249 employees 1 employee $\geq$1000 employees	$\geq$ 500 employees 100 - 499 employees

Source: empirica [7]

FIGURE 7:
The Structural Fit of Supply and Demand
for Teleworking Arrangements

In the short run, telework can be expected to spread to jobs requiring low qualifications (e.g. typing, data entry) and to areas such as data processing where higher qualifications are required. However, it appears that in the longer run, telework in Europe is going to be utilized mainly within more complex and

higher qualified occupations. The developments are likely to take place against the background of various trends underway or emerging:
- automation of routine office tasks,
- increasing trend towards organizational and regional decentralization accompanied by a transfer of competence and responsibility to peripherally located employees,
- implementation and use of more sophisticated IT that support the performance of more complex tasks,
- trend towards an increasing use of IT among professionals,
- off-shore office work (shift of routine office work performance into Third-World countries) [21].

Today, however, management inertia appears to be a major factor inhibiting interest in telework by the majority of decision makers [3,5,19]. The main reasons for not using telework identified within the Decision Maker Survey are:
o no need to change from current situation,
o cost of hardware, software and networks,
o organizational complexity and effort (lack of supervision and control) of personal and performance level.

It seems that the present organizational structures and managerial attitudes and inertia form a major barrier to the spread of telework, though it is not clear whether this is due to the lack of familiarity and experience with telework of decision makers or reflects managerial opposition and difficulties.

5. OPERATIONAL TELEWORK - REASONS AND MOTIVATION OF COMPANY DECISION MAKERS AND INDIVIDUALS FOR STARTING AND TAKING UP TELEWORK

The motivation of company decision makers and individuals for starting and taking up telework were investigated by means of case studies in 14 British and German companies practising telework in 1987. Company managers as well as 119 teleworkers were interviewed [6].

To the author's knowledge, these case studies constitute the most complete survey of telework to date and are representative of the variety of telework activities in Europe.

5.1 Reasons for Telework Inception from the Employer's Perspective

Six reasons predominate in motivating companies to set up telework schemes:
1. to cope with work peaks;
2. to retain the scarce skills of individuals unable to work in a conventional office environment;
3. to reduce costs e.g. for overheads, social benefits. Opportunity to increase a company's capacity at minimal expense;
4. to meet employees' desire to regulate their own work;
5. to recruit urgently required and otherwise not available skills.

The reasons cited by companies for introducing telework differ according to the tasks involved. Company managers responsible for telework schemes where typing tasks are performed give the highest ratings for factors such as: coping with work peaks, rationalization, reduction of employee turnover as well as increasing the opportunities for employees to combine work and non-work activities (especially child care) and their flexibility in working hours.

On the other hand, companies involved in more complex work (e.g. consultancy or programming services) mention as their foremost reasons for starting this new work organization: the retention and recruitment of scarce skills, the improved motivation and increased productivity of their employees/suppliers as well as improved ability to cope with varying work loads.

5.2 Reasons for Telework Inception: The Individual's Perspective

The major advantages cited by teleworkers for starting this work arrangement in order of importance are:

	Total	Male	Female
1. Need for flexibility (ability to work when it suits me), (N=103)	91.2%	68.2%	97.4%
2. Child/family care during working hours (ability to combine care of children and other dependents with my work), (N=91)	87.9%	60.0%	93.2%
3. Family demands (ability to meet the demands of the family), (N=109)	71.5%	50.0%	86.7%
4. Lifestyle demands (ability to combine other activities with my work), (N=100)	71.0%	42.1%	78.2%
5. Commuting hassles (travel and commuting time and expense), (N=102)	69.7%	60.9%	71.1%

The most important advantages of telework relate to the individual's desire for greater flexibility of work location and time. Female teleworkers place extraordinarily high ratings on these considerations. It appears that telework, especially electronic homework, could be seen as a feasible option by a considerable number of women in some phases of their life cycle.

Telework schemes often arise solely from changes in the work organization of existing companies towards decentralization. However, it became apparent that a quite different source of telework inception - schemes directly resulting from new business creation - are also significant.

More than one third (34.2%, N= 73) of the teleworkers surveyed indicated that the opportunities telework offers for developing skills of use to set up their own business were important or very important advantages of these arrangements.

6. CLASSIFICATION OF TELEWORK INCEPTION

Two classifications of telework inception and schemes are presented below. Both are based on the results of the case studies undertaken by empirica in 1987 [6]. The first classification contrasts differing company strategies followed in telework inception. The second is based on the initiator of a telework scheme, which may be an organization or an individual.

6.1 Company Strategy Classification

The following figure shows the construction of two ideal types of teleworking arrangement, divided according to two polar strategies - externalization of labour and access to scarce skills - of the organization setting up the telework scheme. Of course, simplification lies in the fact that these motivations or strategies are present to greater or lesser extent in most telework inceptions. Equivalently, the corresponding teleworker characteristics are idealized and simplified.

If the company objective is the externalization of labour to former employees (conversion of fixed costs into variable ones), it results in a predominance of negative features for teleworkers. Positive outcomes prevail where a company utilizes telework for the purpose of retaining or recruiting scarce skills.

With abundant skills, such as typing and routine office tasks, company strategies are increasingly characterized by sub-contracting work out of the central offices into the homes of the former employees (saving overhead expenses), shifting the contractual arrangement and employment status from full-time salaried staff to flexible part-time and freelance work. This results in a small core staff and an increasing, less expensive and more flexible peripheral work force.

In cases where companies need highly qualified and scarce skills, work arrangements, contractual arrangements, etc., are implemented which best suit the needs of the teleworkers. This often involves flexible work arrangements and provision of important informal communication between the employer, the teleworker and core staff. Moreover, these teleworkers are usually significantly better equipped with IT, which facilitates processing of information and electronic communication.

Currently, a segmentation can be observed between low paid female freelance telehomeworkers for whom telework is seen as a feasible work option in a particular phase of their life cycle, and predominantly qualified male occasional teleworkers. For the latter, telework is lucrative; the work arrangement is to their own convenience and best meets their needs. It offers them flexibility with regard to working location and times and/or for starting their own business.

The idealized duality described reflects important observable segmentation of teleworking schemes, which is related to the market power of the teleworkers concerned. In task areas where there is an excess supply of labour, telework tends to result in a cumulation of negative aspects for the corresponding teleworkers. Correspondingly, where demand well exceeds supply of labour, positive aspects tend to cumulate in the teleworker's favour. Telework is beginning to play an increasing role in company strategies aiming at sub-contracting work out of the central organization resulting in a reduction of core staff and an increase in the proportion of less expensive and more flexible decentralized work force.

Externalization of labour	Access to scarce skills
The Teleworker	**The Teleworker**

o female
o 30 - 40 years old
o responsible for child care
o typists, DP-professionals
o market power reduced by one or more of:
  - family commitments
  - immobility
  - low qualifications
	o male, some female
	o 25 - 65 years old
	o not responsible for child care
	o professional (e.g. business consultants, journalists)
	o market power: high

The Employment	**The Employment**
o electronic homework o part-time o freelance basis o low pay	o flexible work arrangement o full-time o self-employed; creation of one's own business; full-time salaried staff o high income
Motivation	**Motivation**
o flexibility in working times o ability to combine child care with work	o flexibility in working times o autonomy, self-dependence o development of skills for setting up a business o loose supervision
Problems	**Problems**
o pensions o pay level o communication between employer, employees and colleagues o variability of work load o promotion opportunities	o clerical support and other office services copying facilities o benefits, perks, pension schemes o contacts with colleagues, o amount of leisure time
Satisfaction: high	**Satisfaction: high**

FIGURE 8:
Telework as a Company Strategy

6.2 Classification by Initiator

The differing ways in which telework schemes originate and the main purpose the principle actor or actors pursue in the inception process have been classified in Figure 9. This gives an overview of the wide range of existing telework schemes and provides a framework to analyze new schemes.

The classification has been applied to more than 70 existing telework projects including our case studies. With the exception of neighbourhood work centres all the forms of telework described earlier can be observed in reality. It appears that neighbourhood work centres today exist only as a theoretical concept.

Companies reorganizing their performance of work utilize a variety of different concepts. Besides the implementation of various forms of electronic homework, another major area of telework application is the establishment of satellite offices. Here, companies shift work to more or less coherent (functional) subunits of the central organization which are located elsewhere. Advantages accrue through:
- proximity to customer or client location;
- proximity to employees' homes or to a suitable 'pool' of labour. This is particularly important where skills urgently required by the organization are in short supply.

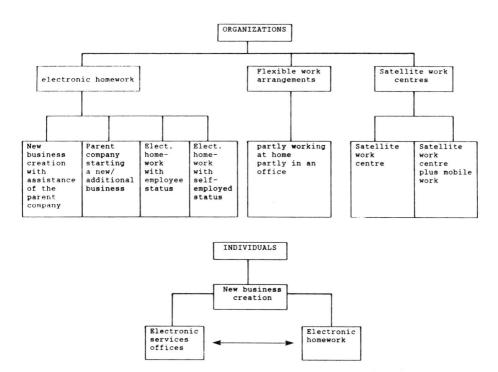

Source: empirica [6]

FIGURE 9:
Classification by Initiator

A further movement towards telework can be seen where existing branch offices of organizations - and here the insurance industry is a typical example - are becoming linked to central sources of information using telecommunications. As links improve, smaller and more widely distributed units become viable. Further delocalization takes place where agents of the organization - who may or may not be employees - are equipped with portable data processing and digital communications equipment, allowing them to work from any location, including their homes. However, in many cases the most important market advantages arise where work can be performed at clients' premises.

Other forms of telework initiated by companies can be described as flexible work arrangements. These provide employees with flexibility in the scheduling of work and in its location. The choice is mainly between the organization's offices and home, though again clients' premises may be an option.

Individuals take up telework in their attempts to tap labour markets by offering their services electronically to potential clients and customers. Often, this is based on new processes and new forms of delivery of products and services made possible by IT. Existing examples cover both individuals working from home and individuals who set up electronic services offices to offer their services to several different firms and clients, using IT, usually in conjunction with a telecommunications network.

In several cases electronic services offices themselves again employ teleworkers working in electronic homework arrangements.

7. CONCLUSIONS

1. The potential for telework implementation, as indicated by awareness of and interest in telework by the workforce and company management is greater than the proportion of organizations and individuals actually utilizing telework.
2. This is, to a significant extent, due to present organizational structures and management attitudes and inertia forming a major barrier to the rapid spread of telework.
3. However, a number of firms throughout Europe have found telework a useful tool to gain comparative advantages over their competitors. Telework allows an easier recruitment and retainment of scarce skills as well as costs reductions and also increases the flexibility of the organization.
4. Telework inception should be seen generally as a response to crises in the work situation faced by an organization or individual. Telework is either the result of
    o attempts by companies to decentralize office work in response to
        - corporate pressures and/or
        - employee pressures (preferences);
    o attempts by individuals to establish electronic services activities as a new business venture, often based on a new form of delivery of products and services and a new form of process;
    o attempts by individuals to continue in paid work despite reduced work ability (disability) or new non-work demands, especially the need to care for a family.
5. Operational telework today is best seen in relation to two polar strategies of the organization setting up the telework

scheme. If the company objective is the externalization of labour to former employees (conversion of fixed costs into variable ones), it results in a predominance of negative features for teleworkers. Positive outcomes prevail where a company utilizes telework for the purpose of retaining or recruiting scarce skills.

6. Company strategies are increasingly characterized by the sub-contracting of work out of the central organization, shifting the balance of contractual arrangements from full-time salaried staff to flexible part-time freelance working. Core staffing is being reduced, and the proportion of less expensive and more flexible decentral work forces is increasing. Telework is beginning to play an increasing role in this process.

7. However, the large number of current telework schemes involving the performance of low qualified tasks and contracted-out (externalized) routine labour by large organizations will only be of temporary duration. It reflects managements' lack of the skills necessary to cope with the reorganization and redesign of companies when introducing IT. To successfully utilize the options IT offers, management will have to shift competence and responsibility to workers on the one hand and use IT to automate routine tasks on the other hand. This will result in the implementation of enlarged and enriched work demanding higher qualifications from those performing office work. Accordingly and at least in the long run, telework is most likely to encompass the performance of mainly qualified, more complex office tasks because of, for example
   - growing automation of routine tasks,
   - increasing work-related use of sophisticated IT among professionals,
   - organizational developments towards decentralization allowing for and finally resulting in job enlargement and job enrichment,
   - better coping of management with new challenges occuring in the frame of managing remote workers.

8. Coupled with changes in values reflecting a growing desire for flexibility which can currently be observed among European populations, telework will spread in the future mainly in the form of flexible work arrangements which meet the needs of both employers and workers best. On the other hand, larger organizations will continue to externalize labour to achieve cost reductions and an increasing flexibility resulting in telework applications like electronic homework and electronic services offices.

9. Today, the proportion of organizations and individuals actually utilizing telework is rather low on the one hand and yet on the other hand the potential for telework seems much higher. Moreover, a large variety of different forms of telework can currently be observed. It is against this background that there seems to be both enormous potential and considerable latitude for designing positive and socially acceptable forms of this new work organization.

Technology policy is also social policy. An initiative should be taken to reach agreement on desirable future perspectives of telework and develop concepts for its use which are socially acceptable, i.e. relate technical innovations to social innovations, prevent negative consequences without depriving those workers who see positive and lucrative opportunities in teleworking.

REFERENCES

[1] Ballerstedt, E., 1985: Telearbeit. In: Die Neue Gesellschaft. Frankfurter Hefte 3, pp. 219-226

[2] Battelle/Integrata, 1982: Informationstechnisch gestützte Heimarbeit. Studie über Auswahl, Eignung und Auswirkungen. Frankfurt, Tübingen

[3] empirica, 1986: Telearbeit - Meinungen und Standpunkte der Sozialpartner und der Erwerbstätigen sowie das Potential dezentraler informationstechnisch gestützter Büroarbeit in Europa, Bonn. Report for the European Foundation for the Improvement of Living and Working Conditions

[4] empirica, 1987: Profiles of the Population Potentially Concerned with Telework - The Supply of Teleworkers. Results of the Employed People Survey (EPS). ESPRIT project 1030, empirica working paper no. 6, Bonn

[5] empirica, 1987: Market Potential for Decentralized Office Work Based on Information Technology - The Demand for Telework. Results of the Decision Maker Survey (DMS). ESPRIT project 1030, empirica working paper no. 7, Bonn

[6] empirica, 1987: A Survey of Teleworkers and Teleworking Companies - Case Studies in Current Telework Environments. ESPRIT project 1030, empirica working paper no. 11, Bonn

[7] empirica, 1987: Overall Report: Potential and Uptake Dynamics of Telework. ESPRIT project 1030. Bonn

[8] FIET 1985: Technology Report No. 1, Geneva

[9] Gordon, G.E., 1987: The Dilemma of Telework: Technology vs. Tradition. In: Korte, W.B./Steinle, W.J./Robinson, S. (Eds.): Telework - Present Situation and Future Development of a New Form of Work Organization. North-Holland, Amsterdam

[10] Heilmann, W., 1986: Art der Tätigkeit als prägender Kontextfaktor der Telearbeit dargestellt am Beispiel der Teleprogrammierung. Presentation at the empirica workshop "The Potential for Decentralized Electronic Working in the Banking, Insurance and Software Industries" on September 18, 1986 in Bonn.

[11] Judkins, P. et al., 1986: Networking in Organizations. The Rank Xerox Experiment

[12] Karcher, H.B., 1984: Büro der Zukunft. Dominanz der Mikros und der Multifunktions-Endgeräte. In: Office Management Nr. 10, pp. 882-887

[13] Köhler, E./Moran, R./Tansey J., 1987: Telework in the European Community: Problems and Potential. In: Korte, W.B./Steinle, W.J./Robinson, S. (Eds.): Telework - Present Situation and Future Development of a New Form of Work Organization. North-Holland, Amsterdam

[14] Korte, W.B./Steinle, W.J., 1985: Telearbeit – Interesse und Akzeptanz. In: Hansen, H.R. (Hrsg.): GI/OCG/ÖGI-Jahrestagung 1985. Wirtschaftsuniversität Wien. Informatik Fachberichte Nr. 108. Springer Verlag Berlin, Heidelberg, New York, Tokio, pp. 959-969

[15] Korte, W.B./Steinle, W.J., 1986: Kultur, Alltagskultur und neue Informations- und Kommunikationstechniken. In: Aus Politik und Zeitgeschichte, B3/86, pp. 26-38

[16] Müllner, W., 1985: Privatisierung des Arbeitsplatzes. Chancen, Risiken und rechtliche Gestaltungsarbeit der Telearbeit. Stuttgart, München, Hannover

[17] Olson, M., 1982: New Information Technology and Organizational Culture. In: MIS Quarterly Special Issue, pp. 71-92

[18] Olson, M./Tasley, R., 1983: Telecommunications and the Changing Definition of the Workplace. New York

[19] Olson, M., 1987: Organizational Barriers to Telework. In: Korte, W.B./Steinle, W.J./Robinson, S. (Eds.): Telework – Present Situation and Future Development of a New Form of Work Organization. North-Holland, Amsterdam

[20] Steinle, W.J., 1987: Dezentrale elektronische Tätigkeiten: Marktpotentiale und Zukunftsperspektiven. In: Gehrmann, F. (Hrsg.): Neue Informations- und Kommunikationstechnologien. Ansätze zur gesellschaftsbezogenen Technologieberichterstattung. Frankfurt/New York

[21] US Congress, Office of Technology Assessment, 1985: Automation of America's Offices. Washington, D.C.

(TELE-) HOMEWORK IN THE FEDERAL REPUBLIC OF GERMANY

HISTORICAL BACKGROUND AND FUTURE PERSPECTIVES FROM A WORKER'S PERSPECTIVE*

Herbert Kubicek, Ulrich Fischer

University of Trier, FB IV
5500 Trier, Federal Republic of Germany

1. SUBJECT AND AIMS

In descriptions of the present political debate about telework in the Federal Republic of Germany it is often said that trade unions completely reject these new forms of work organization despite the fact that a considerable proportion of employers and also employees do want them. Trade Unions are also accused of hindering experiments in which chances and risks of different forms of decentralization of work could be examined.

Taking a closer look it becomes clear that trade unions are only against certain forms of decentralization and legal arrangements, under which the weaker part of the workforce would get even weaker and lose the protection of the collective work legislation and the system of social security. But neither employers nor the ministry or the parliament at federal level are willing to adapt the present work legislation to these risks, which are getting higher through the advances in state sponsored telecommunication networks.

Only in two states (Laender) have legal innovations been introduced and only for workers in the public service, and in one state a proposal for a law on telework has been put forward.

It is our position that experiments with new forms of decentralization of work are useful. But the distribution of economic and social chances and risks should be considered from the start. According to the social norms of the Federal Republic's constitution, technical and organizational innovation should be accompanied by an adaptation of the legal framework so as to secure the protection of the weaker parties on the labour market.

Whereas it may be the case that some aspects may only be investigated during or after experiments, it is also true that many problems concerning basic legal arrangements already can be defined and action could be taken on these in advance.

From this position, we argue that legal arrangements for protecting workers from the greatest social and economic risks could open up the discussion about mutually acceptable forms of decentralization and for socio-technical experiments.

---

* This paper is a short version of an original paper in German. Requests for the longer version should be sent to the authors.

As long as employers and politicians deny any risk and refuse legal innovations which only secure the present level of workers' rights, they create distrust. Instead, an open discussion about risks for workers and about preventive action seems more appropriate.

Therefore, in this paper, we wish to indicate from the perspective of employees or dependent workers ways in which the risks of telework can be avoided and the chances it offers taken up. Firstly, however, we have to establish what risks are involved. In order to provide a discerning analysis of risks, we shall draw a distinction between four organizational forms and four possible legal arrangements (see Table 1).

Table 1: Organizational forms and legal options regarding telework

Legal arrangments of employment / organizational forms of decentralization	regular employment status (out-worker, employee)	loan worker employment	home-worker status	free-lancing/sub-contracting
satellite office/ branch office neighbourhood office computel home				

Applying the worst-case method we focus our analysis of risks on the least favourable cases. With respect to work organization, these occur in the case of isolated telework in the home. This organizational form is examined in terms of the various legal forms of employment. The objection can be made that such an analytical approach presupposes the existance of employer interests that the latter do not actually have. It could be argued that, since the risks are only theoretical constructions, it is not necessary to implement precautionary measures. To counter this criticism, we shall evaluate the historical experience of homework and present evidence of employer interests in each case. In addition, we sketch out goals and ways of achieving a socially acceptable form of telework. We examine intermediate forms of decentralization (particularly regional branches and satellite offices), approaches to an extended form of co-determination, as well as general legislation establishing a legal framework for telework.

In conclusion we consider future prospects and we emphasize that the telecommunications systems themselves can be moulded in various ways. The plans for ISDN in particular are problematized, looking at other risks, too.

2. RISKS INVOLVED IN ELECTRONIC WORK IN THE HOME FROM A WORKERS' PERSPECTIVE

To the extent that the legal relationships binding those working at home to the employing organization are relaxed, telework in one's own home becomes a more attractive business option. From a workers' point of view, however, this also entails greater risks for the individual /2;3;5;16;17;24;25/ (cf. Table 2).

Table 2: Risks for the individual from telework in the home

Legal form of employment	Risks for the individual from telework in the home
Maintenance of employee status	- problems of supervising employment protection regulations (e.g. health and safety) - problems of representing workers' interests - problem of social contact
Loan Worker Employment	In addition: - danger of illegal subcontracting - difficulties in determining the correct rates of pay in each case
Homework Act	In addition: - lower wages - less holiday entitlement - shorter terms of notice - limited rights to employment protection
Freelancing/ sub-contracting	In addition: - cessation of all protection provided unter collective labour law - personal responsibility for social security

Alongside worsening conditions for the individual, it is also conceivable that negative consequences will arise for the system of social security and the organizing power of trade unions. An overview of the collective risks involved in telework in the home is provided in Table 3.

Table 3: Collective risks from telework in the home

System of social security	- lower income - higher contributions - greater recourse to welfare assistance benefits
Organizing strength of trade unions	- problems of the trade union's ability to organize - reduced effectivity of the weapons available for trade union struggle - blocking the demands of workers at the company workplace

3. EMPLOYER INTERESTS IN HOMEWORK

History shows that employers have economic and power political interests in homework in a socio-economic meaning. These interests which are summarized in Table 4. For each of these interests, we specify whether or not it seems plausible that they will again apply in the future.

Apart from the wish to avoid guild regulation, one may assume that all the earlier employer interests in homework remain valid under present conditions.

The historical evidence suggests that the main disadvantages from the employer's point of view are:

a) insufficient opportunity for supervising the production process and exerting discipline in the labour process;

b) insufficient scope for making maximum use of the technical potential for rationalization.

These disadvantages can be overcome now or in the near future, mainly due to the policy on telecommunications. This means that the fears expressed about risks appear to have more substance than the frequently heralded and equally vague visions of new freedoms for all.

Table 4: Employer interests in homework from the workers' perspective

Employer interests in homework	Historical evidence	Could these interests apply in the near future?
Overcoming labour shortages	- Mobilization of homeworkers in the 18th, 19th and 20th centuries in several branches in order to profit from expanding markets /14, p. 202; 23, p. 118ff; 19; p. 1031f/	Yes, for particular skills
Taking advantage of low-wage sectors, including saving on ancillary wage costs	- employing cheap labour found "outside the competitive sphere of the factory" /13, p. 56/ - end of the 19th/beginning of the 20th century: use of homework to avoid payment of social security contributions /20, p. 45/	Yes
Saving on plant and working capital	- use of homework in the 19th century as a result of a high capital demand for plant and material /22, p. 84; 12, p. 363; 1, p. 534; 19, p. 111/	Yes
Adapting production to fluctuations in demand	- major fluctuations in the volume of home-based industrial production in the phase of proto and early industrialization - buffer function of home work in combined production centres towards the end of the 19th and in the 20th century /23, p. 130; 6, p. 304/	Yes

Continuation of Table 4:

Employer interests in homework	Historical evidence	Could these interests apply in the near future?
Neutralizing guild structures	- delegating homework to the non-guild countryside and non-guild urban producers /11, p. 268ff; 26, p. 57ff/ - illegal employment of guild apprentices and journeymen/26, p. 59f/	No
Averting collective opposition	- use of homework to combat trade union influence and undermine strikes in the wake of the growing trade union strength towards the end of the 19th century /10, p. 196f; 9, p. 305/	Yes
Neutralizing employment protection legislation	- use of homework following the spread of employment protection for workers in manufacturing plants towards the end of the 19th century /18, p. 644, 660, 673ff, 695f/	Yes

4. GOALS AND WAYS OF ACHIEVING A SOCIALLY ACCEPTABLE FORM OF TELEWORK

Despite being given little public attention, there are in fact ways of developing telework that avoid the potential disadvantages.

(1)   Organizational options: intermediate decentralization as a positive perspective

Alongside the work performed in the home in certain special circumstances (second workplace in the home; temporary work at home with a guaranteed return to the workplace, e.g. in the case of child-care arrangements), we regard intermediate stages of decentralization as a positive perspective.

There is no objection to satellite offices (i.e. branches of a company situated near workers' homes) if the following conditions are met:

- For effective representation of interests in the area of law relating to the workplace organization it is necessary to have a minimum of 20 employees who are entitled to vote. With fewer employees (a minimum of five) it is only possible to elect a staff spokesman who has fewer rights.

- The functions of branches could be selected in such a way that it is possible to work on all aspects of a job and, where possible, provide responsible customer service and advice, too.

- It must be possible to limit work on VDUs (monitors) to 50% of the working day at every place of work.

- A form of data access and data security must be found which minimizes supervision of individual behaviour as far as possible or at least rules out disadvantages that might result for employees.

In contrast to branches and satellite offices, neighbourhood offices and neighbourhood work centres represent new organizational forms /8; 21/:

- Neighbourhood offices describe spacially integrated office workplaces in residential areas used by workers for various employers. Proximity of workplaces to the home is seen as the decisive criterion here.

- Neighbourhood centres designate facilities designed to combine various activities under one roof which provide cultural (i.e. library) and gastronomic services etc. The primary objective here is to reactivate social contact in suburbs and innercity areas.

If there is any future in these projects, then it lies in the direction of social experiments with cooperatives where specific technical resources are shared. In addition to many other fields of activity, this option might be suitable for such projects as action groups for the unemployed or "alternative businesses", and certainly for organizing individual activities associated with telecommunication facilities. The emergence and spread of such organizational forms is not, however, primarily a technical problem but an economic, social and political one.

(2) Extended rights of co-determination

A legal right to co-determination on the creation of teleworkplaces would provide an opportunity to develop regulations in relation to the specific situation of the company and its employees and to root these in company agreements. These agreements, for their part, could be developed further on the basis of mutual consent. All in all, then, this represents a very flexible means of achieving socially acceptable forms of telework.

At present only two states in the FRG have staff representation acts explicitly providing for equivalent rights of co-determination:

- Under the Hesse State Staff Representation Act co-determination is related to the assignment and privatization of work or tasks previously performed by employees of the agency. In addition, co-determination also arises in connection with new working methods, of which electronic telework is one.

- In the North Rhine Westphalia State Staff Representation Act a distinction is drawn between the "relocation of workplaces in favour of homework being performed on technical equipment" and the "transfer of company tasks normally performed by employees on the premises to private persons or businesses on a permanent basis."

Similar regulations are also called for in SPD proposals and in a German Trade Union Federation draft for the re-enactment of the Employees' Representation Act.

(3) The Hamburg draft of a telework law

Proposals for a combination of extended industrial democracy and a legal framework for telework was submitted by the Work, Youth and Social Affairs Authority of the Hamburg Senate. The draft comes nearest to the requirements presented from a workers' point of view in this paper as a whole. Summing up, the draft contains the following regulative elements /3, p. 558/:

- a ban on the employment of teleworkers under revised conditions of employment, i.e. the legal prevention of contractual relationships for homeworking, free-lancing and subcontracting etc.;

- safeguarding the legal relationship binding the teleworker to an employing agency;

- a ban on teleworkplaces in the home, subject to reservations permitting their creation in special circumstances (i.e. in the case of the severely handicapped for whom allocation of work on the company's premises poses particular difficulties; employees with seriously ill family dependents living at home) and, as a further condition, subject to agreements by a committee made up equally of representatives from the leading organizations of employers and employees;

- an arrangement permitting the establishment of teleworkplaces in neighbourhood and satellite offices only on condition that there are enough employees to form viable staff councils;

- a general ban on discrimination against teleworkers (vis-a-vis employees working on the company's premises) that is specified in terms of financial remuneration and rules out any loss of social benefits for employees;

- a regulation obliging the employer to pay the teleworker lump-sum compensation for wear and tear of furniture and fixtures in the homeworkplace, for the share of rent or lease payments going on teleworkspace, or for the corresponding proportion of energy, heating and water costs;

- arrangements limiting working hours, in particular preventing telework between 8 p.m. and 6 a.m. and restricting overtime to 2 hours per week;

- a ban on the conclusion of agreements for labour to be hired on an on-call basis;

- various regulations to secure satisfactory working conditions from the standpoint of ergonomics and occupational medicine, such as a stipulation on regular medical check-ups and legislation supporting a claim on the part of a teleworker to be allocated equivalent work tasks or acceptable occupational activities if he is no longer able to work on a video disply unit for medical reasons;

- an arrangement establishing that periods in which work systems are out of order are to be paid as working hours;

- regulations obliging the employer to give teleworkers notice of company vacancies and give them preferential treatment in such matters as appointing staff for posts at the company's premises;

- a ban on transfering the costs for installation, maintenance and servicing of teleworkplaces, declaring any contrary agreements to be invalid,

- a limitation on liability in the case of loss of or damage to company equipment and material by intent or gross negligence, as well as (in such cases as damage caused by children) a limitation of liability against gross violation of supervisory duties; and

- legislation securing a minimum additional holiday of two days for teleworkers, a directory provision on the establishment of mixed workplaces, notification (posting) of regulations, and the settlement of regulatory offences which envisage big fines in the case of employers contravening the aforementioned directives and prohibitions.

We are not saying that this is the final and optimal set of regulations. Rather, this is a starting point for a detailed discussion which, however, has not really taken offe in the two years since the presentation of this draft.

5. ON THE RELATIONSHIP BETWEEN TECHNICAL AND SOCIAL INNOVATIONS

At present the limited spread of telework has to do with such factors as the costs of technical facilities, including the transmission costs. Through the conversion of the telephone system to the ISDN standard more favourable technical and economic conditions would be created. ISDN is being given massive support, both at the national and European Community levels. This support is largely a result of growth-oriented policy goals. However, coordinating efforts to achieve a unified standardization and schedules is not matched by any comparable efforts to avoid and limit the social risks involved. This means that, among other things, ISDN would lead to the creation of a more favourable situation in which employers can dodge employment regulations currently in force and avoid their financial contribution to the social security systems. Yet even more problems would be posed by customer self-service telecommunications (telebanking, teleshopping etc.), that are also becoming a more attractive option. This option actually involves the minute-for-minute transference of data-input work to the customer, i.e. the home or the public terminal as a "customer workplace". This is a type of rationalization for which present rights of co-determination and technology agreements were not meant and which therefore is not covered by these regulations.

If it were really a question of promoting social innovations, it would not be sufficient to take belated social policy measures to limit and compensate for risks. On the contrary, one would have to ask how much decentralization is desired in social policy and which telecommunications infrastructure meets this aim. At the present time, many people appear to believe that the establishment of a comprehensive integrated computer system is free from the influence of interests. HEDBERG and JOHANSSON-HEDBERG have drawn attention to the correspondence between social and technical structures /7/:

> " The choice between future computerization at the home level (including homework) and a computerization at the institutional level (including central workplaces) is an example of an incredibly important strategic choice for the future development of the imformation society. Computerization at the home level is linked to homework, home-banking, home-shopping, computer-assisted teaching, and electronic post, newspaper or financial transactions, as well as greater media availability in the home. Computerization at the institutional level (central workplaces, local decision-making centres, neighbourhood centres) is more compatible with present institutions in society, such as the postal service, banks, schools, day-centres, libraries or the press."/21, p. 23/

Expressing this more precisely, we can say that the politically motivated conversion of the telephone system to the ISDN standard and the favouring of data and text transmission signifies preferential support for flexibilization and individualization strategies currently being pursued by conservative politicians and businesses. In the individualized society, the influence of many traditional collective institutions would be undermined. Free, direct inter-personal communication would be replaced by interaction via rented communication lines and the consumption of prerecorded information and entertainment. Economic growth o.k., but at what price? /15/

If one would like a kind of society characterized by social justice and security and a lively, multi-facetted culture, then one also has to create a different telecommunications infrastructure. From this perspective it no longer appears desirable, e.g., to provide terminals for rapid data and text transmission available at cheap prices in every home. Rather, it would be sufficient, to provide corresponding facilities for businesses as well as public access facilities for the occasional user. Right from the beginning this entails a smaller volume of investment in building up the network and a smaller market for consumer appliances. And this would then pose the question as to whether it would be better to consolidate the existing special network, instead of converting the telephone system, or whether pricing policy should give priority to cheapening telephone calls rather than data transmission /16/.

At present those who have an interest in turnover, profits and markets, as well as in the removal of employee rights, exercise a determining influence on telecommunications and technology policy of the EC and national governments. However, technology policy is also social policy. Our major worry is that while the social policy makers are reaching agreements on desirable future perspectives for telework, policy-makers in the economics and technology sector are taking concrete steps which promote developments leading in a completely different direction.

6. REFERENCES

/1/ Baar, L.: Probleme der industriellen Revolution in großstädtischen Industriezentren. Das Berliner Beispiel. In: Wirtschafts- und Sozialwissenschaftliche Probleme der Frühen Industrialisierung, edited by W. Fischer (Berlin 1968), pp. 529-534

/2/ Bahl-Benker, A.: Elektronische Heimarbeit - die "schöne neue Arbeitswelt?" In: Die Mitbestimmung 12/1983, pp. 572-576

/3/ Beck, T.: Elektronische Fernarbeit und Arbeitsrecht - Rechtspolitische Überlegungen zu Entwicklungen, Regelungsbedarf und Regelungsmöglichkeiten. In: WSI-Mitteilungen, 9/1985, pp. 550-558

/4/ Bittmann, K.: Hausindustrie und Heimarbeit im Großherzogtum Baden zu Anfang des XX. Jahrhunderts. Karlsruhe 1907

/5/ Goldmann, M./Richter, G.: Teleheimarbeiterinnen in der Satzerstellung/Texterfassung für die Druckindustrie. Ergebnisse einer Frauenbefragung. Dortmund 1986

/6/ Grunow: Die Solinger Industrie. Eine wirtschaftliche Studie. In: Schriften des Vereins für Socialpolitik, Bd. 88, Leipzig 1900, pp. 269-310

/7/ Hedberg, B./Hedberg-Johansson, B.: Kroneberg: 2000 - om liv och arbete i informationssamhället. Stockholm 1983 (Arbetslivscentrum)

/8/ Holzer, C.: Die Nachbarschaftszentrale - ein Beispiel zur Technikgestaltbarkeit? Diplomarbeit, Karlsruhe 1984

/9/ Jaffe, E.: Hausindustrie und Fabrikbetrieb in der deutschen Cigarrenfabrikation. In: Schriften des Vereins für Socialpolitik, Bd. 86, Leipzig 1899, pp. 279-341

/10/ Kieferitzky, E.: Die Formen der Hausindustrie in Köln. In: Schriften des Vereins für Socialpolitik, Bd. 86, Leipzig 1899

/11/ Kisch, H.: Das Erbe des Mittelalters, ein Hemmnis wirtschaftlicher Entwicklung: Aachens Tuchgewerbe vor 1790. In: Rheinische Vierteljahresblätter, 1965, pp. 253-308

/12/ Kisch, H.: Das Textilgewerbe in Schlesien und im Rheinland: eine vergleichende Studie zur Industrialisierung. In: Kriedtke, P./Medick, H./Schlumbohm, J.: Industrialisierung vor der Industrialisierung. Gewerbliche Warenproduktion auf dem Land in der Formationsperiode des Kapitalismus, Göttingen 1977, pp. 350-386

/13/ Koch, H.: Die deutsche Hausindustrie, Mönchengladbach 1905

/14/ Krüger, H.: Zur Geschichte der Manufakturen und der Manufakturarbeiter in Preußen. Die mittleren Provinzen in der zweiten Hälfte des 18. Jahrhunderts, Berlin 1958

/15/ Kubicek, H.: Die sogenannte Informationsgesellschaft. Neue Informations- und Kommunikationstechniken als Instrument konservativer Gesellschaftsveränderung. In: Altvater, E./Baethge, M. u.a.: Arbeit 2000. Über die Zukunft der Arbeitsgesellschaft, Hamburg 1985, pp. 76-109

/16/ Kubicek, H./Rolf, A.: Mikropolis. Mit Computernetzen in die "Informationsgesellschaft", Hamburg 1985

/17/ Lohmar, U.: Die neue Heimarbeit. In: Gitter, W. u.a.: Telearbeit. Beiträge zur Gesellschafts- und Bildungspolitik Nr. 109. Institut der deutschen Wirtschaft, Köln 1985, pp. 7-25

/18/ Meerwarth, R.: Untersuchungen über die Hausindustrie in Deutschland. In: Schriften der Gesellschaft für Soziale Reform, 2. Band, Jena 1907, pp. 627-697

/19/ Protokoll der Verhandlungen des ersten Allgemeinen Heimarbeitsschutzkongresses, Berlin 1904

/20/ Reinhard, O.: Die württembergische Trikot-Industrie mit spezieller Berücksichtigung der Heimarbeit in den Bezirken Stuttgart (Stadt und Land) und Balingen. In: Schriften des Vereins für Socialpolitik, Bd. 84, Leipzig 1899, pp. 1-77

/21/ Simoleit, J.: Sozialpolitische und beschäftigungswirksame Formen der Anwendung neuer Informations- und Kommunikationstechnologien, Materialien aus der Kooperationsarbeit der Kooperationsstelle Hamburg, Nr. 6, Hamburg 1985

/22/ Slawinger, G.: Die Manufaktur in Kurbayern. Die Anfänge der großgewerblichen Entwicklung in der Übergangsepoche vom Merkantilismus zum Liberalismus 1740-1833, Diss. München 1966

/23/ Thun, A.: Die Crefelder Seidenindustrie und die Crisis, Jahrbuch für Gesetzgebung, Verwaltung und Volkswirtschaft im Deutschen Reich, 1879, pp. 111-143

/24/ Ulber, J.: Heimarbeit, neue Technologien und betriebliche Interessenvertretung, Arbeitsrecht im Betrieb, 2/1985, pp. 22-28

/25/ Wedde, P.: Telearbeit und Arbeitsrecht. Schutz der Beschäftigten und Handlungsmöglichkeiten des Betriebsrates, Köln 1986

/26/ Zwahr, H.: Zur Konstituierung des Proletariats als Klasse. Strukturuntersuchung über das Leipziger Proletariat während der industriellen Revolution, Berlin 1978

AUTONOMY, TELEWORK AND EMERGING CULTURAL VALUES

Gerard Blanc

Research Associate, Association Internationale Futuribles
55 rue de Varenne, 75007 Paris, France

1. TELEWORK: MORE HASTE, LESS SPEED?

A few years ago, telework was heralded as the next revolution in working and domestic life. Today, it seems to catch on in North America while Europe lags well behind. Yet telework is advancing in various situations, activities, businesses, together with innovative work organization and/or localization. Formal and referenced experiments by big corporations - sometimes at relatively large scale - mix with numerous unnoticed, isolated, individual practices. More than three hundred different trades where telework is practised have been registered in the United States. Some investigators have tried to count teleworkers, but none has conducted an exhaustive survey and it is still impossible to go beyond guesses.

Future prospects for telework development seem quite good, especially if it is taken in its broadest meaning which includes all forms of decentralized organization in which work is done at a distance from its "usual" location.

In a recent survey [1], 240 social sciences experts from OECD countries consulted on socio-cultural trends until the year 2,005 expressed nearly unanimously that telework will develop. More generally, the rise of communication technology will bring about major changes in the spatial distribution of work and housing. No need being an expert to feel this trend. In a sample survey of French white and blue collars, managers, middle managers and graduate students, conducted in 1984 by IPSOS and the European Business School [2], nearly half of each social group members agreed that "telematics and bureaucratics will allow everybody to work from home"; between 37 and 52 per cent of them thought it was a desirable option. In the same way, more than half of the sample and 70 per cent of the graduate students considered quite possible that "the corporation will not be a place where people meet and spend a large part of their time; it will be an organized community of people working in different places and linked together via computers"; a slightly smaller proportion of them thought it desirable.

In Europe, telework does not seem to thrive despite these positive assessments and despite the implementation of its technological prerequisites. Enthusiastic promoters of telework may have gone a little too fast, and may have neglected socio-cultural factors which are inhibiting its growth. Such a reorganization of work, be it only its location, would be a major social transformation already. During the French public transportation strikes in France, combined with bad weather conditions, it has been observed how such changes can be triggered by exceptional circumstances: some banks invited their employees, if they could not go to their office, to work in the closest agency. This innovation may be seen as a first step towards a more rational and less costly distribution of work places and housing, an opening towards more control on their work organization by employees? It could also be the beginning of a public concern on the issue.

Until now, all studies on telework have shown that the transformations it may bring about in the long run make up a deep cultural change. As it happened for many other innovations across history, it is, at the beginning, mostly found among the elite of the society. Actual practitioners of telework are to be found among managers, professionals and workers with skills that are in high demand. At the same time, telework development is highly dependent on the progressive adoption of associated socio-cultural values among the society at large: "small is beautiful", autonomy and self-discipline, cooperation, freedom to organize one's time, intertwining of professional and family life, "back at home" movement, etc. which are examined in this presentation.

2. THE EVOLUTION OF WORK

In order to understand its historical perspective and grasp its dynamics, telework should be examined in the context of long-term processes, which are slowly transforming the world of business and the environment of work. The size, location and organization of business have witnessed more changes in the last fifteen years than during the previous half-century.

While big corporations continue to grow, the nineteen-eighties have seen the emergence of various applications of the "small is beautiful" paradigm. A new generation of managers has lost faith in the virtues of bigness; breaking down large conglomerates, as it happened with ATT, might be the way to improve their competitiveness on the world market. Within companies, co-ordinated small "independent" production units have gained importance in many sectors; for instance, semi-autonomous work groups have demonstrated their feasibility and benefits in car factories. Sub-contracting has grown in all sectors of the economic and it has now reached tertiary activities like typing, accounting, computer analysis or report publishing. French experts consider that subcontracted remote services via telematics will become a growth sector and a major direction of telework development [3].

Tele-informatics, computer programs run at a distance, is increasing the possibility of creating decentralized businesses, which will be able to work with employees spread all over the country. A good case in point is Occitel, a small employee-owned telematics service company, whose members live in various places in the south of France and work together in team or as independent consultants [3]. Another evidence of these changes is the number of businesses which list the home address of their owner as their place of activity; although no hard statistics are available, the United States Chamber of Commerce reported they were ten million in 1984.

From taylorism to self-management, many schools and ideas about business organization and management have raised and declined during the twentieth century. Innovative methods have been experimented with and are slowly introduced into daily operations.

Profit-sharing, workers' participation to decisions have been much talked about in French political speeches, their implementation has remained marginal, except in the service sector, where the proportion of professionals is the highest, e.g. consulting. In a similar way, job-sharing has faced psychological obstacles: when several workers do the same tasks and share the workload, they must accept to be less individualistic, and more cooperative, a behaviour more in the tradition of the East than the West.

There is also a general move of corporate organization away from hierarchical patterns. External supervision through span of control, formal hierarchies of authority and narrow definitions of jobs are being replaced by other principles like professionalism, lower-level decision making or network patterns of communications. Control and coordination are more and more carried on through manipulation of information rather than physical presence.

3. TASK CONTROL AND SUPERVISION

Practitioners of telework are all impressed by the extent to which it upsets hierarchical patterns and relationships. This is a major reason why managers are often reluctant or sceptical towards it. They object to difficulties about controlling and supervising their subordinates' tasks, both quantitively and qualitively. For long managerial staff has insisted more upon respect of formal procedures and checking of employees' attendance than upon actual efficiency. It is an economic nonsense that those means of control become an end in itself, especially for tertiary activities; their output is not measured only by quantity, and a strict 9 to 5 workday is usually no longer critical.

When the employees are working alone in remote places, not visible from their direct managers, the ways of controlling and evaluating their activities have to be reconsidered. Supervisors have to devise new procedures, new regulation tools, for instance by defining intermediate steps more easily accountable for; completing a project or writing a report is split into short phases, each with precise objectives and specific time-spans. Practical experiences show that everybody can benefit from this change. Managers at Control Data Corp., for instance, record that teleworkers involved in the company's "alternative work sites" program have improved their weekly reports giving more details, while they learned how to plan their activities by themselves more efficiently [4].

Managers are thus led to abandon the myth of "being there" or "showing up" as a criterion for efficiency. Duration of work loses its meaning as attendance in the office becomes no longer necessary for actual work. Louis Mertes, vice-president and general manager of systems at Continental Illinois, one of the first American companies to start telework on a large scale, noticed that "when paper was the exclusive medium, the boss never knew whether a memorandum had been drafted in the office, aboard a commuter train or at a kitchen table after the late, late show" [5]. And he did not care. Why should micro-computing make him more curious? The long run trend will be shaped by choices between emphasizing efficiency and output quality - wherever the work is done - and maintaining symbols of hierarchy or authority.

However, telework is not suitable for everyone. Commitment to its practice depends on many individual features and choices. Corporations which have started telework experiments have carefully selected their candidates, who should meet both professional requirements - skills, ability to work independently, seniority, trustworthiness - and strong will.

Volunteers may find in the experiment an opportunity for a promotion, a personal renewal, a challenge or the test of a new technology. Some, previously disenchanted with the slow pace of corporate changes, take the opportunity to give up routine tasks. Others who hold a favourable power position because of their personal skills, may have had the initiative of starting the experiment.

What can trigger this decision to break down conventional working patterns and norms? It may be personal events, like moving off, the birth of a child, a relative's serious disease, a temporary disablement, as well as life choices. It may be a strong will to devote more time to their family, children or hobby. It may be also the priority they give to the choice of their place in life, place of work. But despite these happy or unhappy circumstances, they wanted to keep on working. These personal motivations give them the impetus to succeed, and the necessary strength. All were people who did not accept to follow a conventional career path, entirely determined in advance, and who were ready to make sharp decisions regarding themselves and their lifestyle.

4. AUTONOMY AND SELF-DISCIPLINE

Taking a larger responsibility for one's work organization is not new for many professionals who were already used to work at home from time to time. They just increase their working time outside their office. Those who were unpractised in working at home soon discover the need for a personal discipline. Everybody cannot say at once, like FIL's account manager: "whether my office is at home or not, the moment I walk into it, I'm at work" [6].

For excutives or professionals who were heavily dependent on their secretaries for administrative or typing tasks, the use of a micro-computer, and especially word-processing software, appear as a relief and a welcome feeling of autonomy. Even if it means more work in the beginning before they become accustomed to it. All members of the BURSENS project at the School of Public Administration of Montreal University mentioned this strongly positive transformation [7].

Increase in autonomy has a reverse side: working from home requires a high degree of self-motivation, interest and self-discipline. The teleworker needs to prepare his work better, he has also to get rid of habits fostered by his subordinates' permanent availability. In that respect he can be helped by his micro-computing or telecommunication tools, which are ready for use at any time of day or night. Some find it hard to avoid "workaholism", the excitement of the micro-computer which makes them spend hours and hours in front of it.

Self-control and self-management of one's time-table seems the major embodiment of this increase of autonomy that most teleworkers have felt. The major change in temporal patterns of activities is the suppression or large decrease in commuting. In France, working at home can save half an hour in the morning and half an hour in the evening on the average, but it also spares tiredness, stress and waiting times.

A large number of people wish to take control over the organization of their daily life. They would like to select by themselves which are their working hours and which are their leisure hours, in order to keep a balance between quiet and hectic times or to remain available for their family or friends. These expectations, to which telework can contribute, were well identified in a 1983 survey by the CREDOC: one third of salaried workers wanted more freedom to arrange their own schedule over the week or the month; one fifth of them called for a better account of their family constraints when planning their working hours. A report to the French population and Family Council presented in February 1987 has recently proposed a greater flexibility of work organization in order to adjust work and family life [8].

Office hours do not usually take into account sleeping hours, or times of best intellectual or physical fitness, which vary from people to people, or over the year. Many tertiary activities do not require any synchronization of working hours; more, forcing everybody to begin and end at the same time may well lower the efficiency of the whole. Telework can thus bring flexible schedules, more suited to the diversity of individual biorhythms. They lead to a wide range of practices, for physiological as well as social, economic or family reasons. Typical patterns of activities in the day of a teleworker have not yet been deeply studied. Neither do we know how the time saved on commuting is spent: more domestic chores, especially for women telecommuters, physical or cultural activities, etc.?

It is clear that these teleworkers were tired of wasting time commuting, that they wanted to spend more time with their family, to take a more active role in raising their children. They were ready to stay long at home, even if these supplementary hours are spent working. This is a trend which can be seen in many emerging features of the society. It is not actually new, it is more the revival of what happened before the industrial revolution, when people worked at home and did not commute to work.

5. WORKING AT HOME

Telework at home emphasizes another specific feature of this trend, the narrowing of a gap between professional life and family life.

The split between life at work and life out-of-work became a social standard only during the XIXth century. Nothing proves that it is a psychological necessity and that mingling working activities with family life may upset people. Nothing proves that a majority of households want to keep strongly apart. On the contrary, there is "a will to integrate work and daily life, to overthrow the border between them" among the French "new middle classes" [9].

Surveys conducted in the United States and Canada show that telework has rather been an opportunity to improve family relationships between husband and wife or parents and children. In

Europe, some risks of family conflicts may arise because living spaces are much smaller than in North America; but a tiny flat might be a strong deterrence against the desire to try telework at home.

Moreover, the border between family and professional life is difficult to define and varies according to psychological and educational features. Children tend to reproduce their parents' feelings and behaviour towards this distinction. In that respect, many teleworkers feel that it is important that their children see them at work, because most of the time, children do not know and do not understand what going to work does mean.

The family environment plays a decisive part in the success - or the failure - of telework at home, while this new practice can radically upset family life. The members of the family must accept the teleworker's attendance at home, and simultaneously avoid to disturb him at any occasion. It is a matter of education and habit which should decrease in the long run.

Most teleworkers do not find any objection to the fading out between professional and private life. On the contrary, this practice yields many advantages: the home environment offers better working conditions; it is easier to remain relaxed, to maintain undisturbed working conditions or to prevent tiredness by a break or snack. Teleworkers do insist that they work longer, with a better concentration, on a particular task than if they were in their office. However, a few practitioners of telework feel disturbed by domestic disruptions and have problems to explain to their family that they are at home in order to work.

Finding the right place for a computer in a home generally appears less easy than expected. This issue raises up various solutions: the living room, beside the TV set, the spare bedroom, the basement, the attic, etc. A deeper concern is the ambiguous perception of the computer by family members: is it a working tool, a leisure instrument, a piece of furniture or an ugly machine [10]?

The computer changes the function of the room where it is located; overlap of activities and confusion may occur. Notions of private and public space become jammed because on the one hand, the computer should be put in a private place for the sake of its user's quietness, and on the other hand, it should remain available for visitors. This issue is not specific to telework, but it becomes more sensitive in this case.

Remote work will also induce an evolution of architectural design, as the telephone did in French flats, by moving progressively from the corridor to the living room, then the bedroom and the workroom. In his study of the future of computer terminals in the home, Yves Gassot [11] has noticed a high correlation between telework at home and the enrichment of housing functions. This phenomenon should lead to the extension of working rooms, a demand for larger flats and for more flexible room design, as well as a renewal of the market for apartments with workrooms.

These conclusions meet the scenarios imagined by the authors of a société digitale [12]. The desire to reduce commuting and travel leads to leave a single-use idea of housing for a multi-functional notion. The house becomes again - as it was before the industrial revolution - a "complete place" where its inhabitants work, dwell, spend their leisure time or order their shopping. Autonomous activities in the home - repairs, crafts, indoor vegetable growing, etc. - are becoming more and more fashionable, which witnesses that home is becoming again a place for economic production. Many economic, cultural, technical indicators confirm this "back to home" trend: sales of domestic appliances, access to house ownership, development of homes services and leisure activities, revival of family's virtues and many more [13].

6. ISOLATION, AN IRRELEVANT PROBLEM

Nowadays, sociologists are quite worried by increasing feelings of isolation and loneliness in industrial societies. They consider that isolation could be the worst consequence brought about by telework. It is the most frequently mentioned in articles and reports on the subject. During a presentation of telework at home by a French TV network in the end of 1985, loneliness seemed to be the only one aspect of telework to be worth debating. But actual evidences are rather scarce and what is the meaning of this obscure and ill-defined notion of isolation?

Strong warnings have been given against the risks caused by withdrawal within the family circle and the suppression of social interactions within the workplace, which would dissolve the social fabric. This position assumes that the workplace is the only one place for social encounters or gatherings, which still needs to be proved.

No consensus emerges whether or not people who do not work in an office miss the companionship of their colleagues. A few women teleworkers have mentioned that the telephone is not enough to meet their needs for social interactions. But many others see the isolation as a benefit; they did not appreciate these aspects of office life like endless chats or rattling typewriters and were disturbed or irritated by a mere presence around them.

In a recent poll, one fourth of the surveyed sample denied that the office was their favourite place for social relationships. Their opinion as regards working at home was highly conditioned by the way the question was formulated: asked if they would like to work at home, without any other precision, seven per cent answered yes; but if the question was given a futurist or voluntarist flavour, with a possibility for choices, between 20 and 30 per cent gave a positive answer [14].

A distinction should be made between social isolation and professional isolation. In the second instance, telework can often re-create conditions necessary for team work, and provide with this part of communications which can be mediatized. Technology prevents co-workers from the obligation of being in the same place at the same time in order to share information, exchange advices, elaborate plans, give their opinion about reports, prepare a meeting or a decision, and coordinate everybody's activities. Telecommunications and telematics tools keep coordination, cohesiveness and informal relations within the team. Studies of teleconferences have shown that these means of communication are best suited for people who are already known to each other and who are involved in cooperative discussions. Conflictual situations requiring persuasion and negotiations can be handled only by face-to-face discussions [15].

Telex operators, truck or taxi drivers have been using citizen's band radio since several years; they communicate although out of sight, sometimes they have never seen each other physically. This practice has been naturally extended to users of electronic mail and computer-aided teleconferences. At Control Data, for instance, several teleworkers have joined a company "electronic community" to restore informal exchanges and office gossips. But how far can these particular human relations go?

7. MEDIATIZED RELATIONS: THE PROS AND CONS

Mediatized contacts are actually older than usually assumed and did not start with telecommunications. Wearing spectacles already introduces some kind of mediatization, since it creates a difference between the image you perceive and the image you receive: the face-to-face is already intermediatized. But there is a change in nature as far as the difference between physical presence and its electronic simulation is concerned. Electronically-mediated communication is the instrumentalized form of communication, in the same way that motorized transportation is the instrumentalized form of moving [16].

Contradictory points of views are given regarding the potential effects of this phenomenon. On the one hand, psychologists have noticed that warm feelings still can be transmitted through the telephone; some people even express themselves better, in a more open and sincere way than during a face-to-face conversation. Optimists hope that present and future means of telecommunications will be able to carry this mysterious spark which gives its wit to social and human interactions. Mediatized contacts, thanks to portable micro-computers and mobile telephones, for instance, make the old dream of ubiquity become true - at least partially.

On the other hand, the increase in mediatized relations is not completely harmless. The American anthropologist Edward Hall has shown how human societies use space, the way they feel and design it, to organize and distinguish themselves, take their bearings and communicate [17]. To believe that mediatized relations wipe out distance is an illusion, all the more misleading as distances still do exist and play a part for sorting and grading information, work out rational judgements or set up cooperative processes.

There is no need to go back to the dramatic description of the collapse of a wired society that E.M. Forster gave in his 1909 short story **The machine stops** [18]. The potential dangers of an increase in mediatized relations have been emphasized many times, especially when breakdowns recall the fragility of huge technical systems. As far as telework is concerned, it should be noticed that one of the first extensive reports on the subject submitted in 1970 to the French Land Planning Directorate (DATAR) already warned that "it is no longer an utopia before any human relation, including the most intimate ones, can have its artificial extension" [19].

In any case, telework is deeply connected to this issue; it might well reinforce a trend towards increased mediatized relationships which finds various expressions within today's industrial societies. But teleworkers' actual experiences also show the opposite trend: an increase of direct, close contacts. They visit their friends and neighbours much more than before, provided some preliminary rules are set up, like the respect of their working hours - at home. They also want to become involved in local politics or social activities. In France, evidence for this revival can be found in the growth of non-profit citizens' associations ("associations loi de 1901"), which increased from 20,000 in 1967 to 40,000 in 1980, while their membership became younger and reached one third the French population, compared to one fourth in the 1960's. These associations play an increasing role in local life, especially in small towns and rural areas [20].

These remarks corroborate Alvin Toffler's vision in The third wave where he foresaw that computers and telecommunications would help us develop new community feelings, that local meetings and direct contacts would become fashionable again. This "telecommunity" would strengthen collective stability; human emotional relations would become less superficial. In an interview given in 1984 to the French computer weekly Le Monde Informatique [21], Toffler made his point clear that people who work at home may wish to go out to the restaurant, the theatre or a local meeting at the end of the day; all things they would not do after a day of work outside.

8. A CASE STUDY: THE ILE-DE-FRANCE REGION

This short presentation gives a rather impressionistic picture of the socio-cultural background in which telework is growing. The whole landscape is still an incomplete puzzle, a maze of social transformations, claims for various types of freedom, critics of the existing order, confused tries and innovations of all kinds. These trends should not be considered separately, because their combinations and interactions are at least as important as their individual features.

In order to carry on the discussion on a more concrete basis, it is interesting to observe what could happen at a regional level by the year 2,000. In France, the area around Paris, the Ile-de-France administrative region, seems to be best suited to accept the mutation implied by the extension of telework. About seventy variables of various nature (technical, economic, demographic, political, etc.) which shape the future of telework in the region have been selected and processed through a structural analysis exercise [22]. This method allows to detect the most basic variables, taking into account the many feedback loops, positive or negative, which connect them.

The major conclusions of this research are summarized in the table below. Telework appears to be no utopia at all and should be considered seriously. By the year 2,000, a significant proportion of the work force in the region, at various levels of sales, consulting, customer services, programming, research or administration, could work at home or in neighbourhood centres, two to four days a week. Telework would benefit individuals and communities alike, alleviate commuting costs and partially solve transporation issues, although it could not, by itself, suppress geographical discrepancies between areas where people go to work and areas where they live. Many cultural transformations would emerge, some expected, some unexpected.

However, contrary trends should not be neglected. Some of the most dynamic and affluent groups within French society, for instance, tend to spend less and less time at home, and are not willing to work there. Resistance by management and employees against breaking off old habits may be so strong that only a highly motivated minority would try [23]. Unemployment brings about different attitudes among young French adults. Some of them follow a strategy of amenability, conformism and "everyone for himself". Others want to create their own jobs, and entrepreneurship is presently thriving. In the long run, those will naturally organize their work to their likings and telework might well be a desirable option.

THE FUTURE OF TELEWORK IN THE ILE-DE-FRANCE REGION

Major incentives for telework

1. Jobs are spatially balanced over the whole region, despite high local employment discrepancies (700,000 more jobs than working people in Paris itself).
2. Distribution of dwellings follows a centrifugal pattern: since the nineteen-fifties, couples and families have settled farther and farther from Paris in remote suburbs, where houses are larger and every other household owns his house or apartment while jobs remained mostly in Paris.
3. Since long, transportation issues have been a nightmare for the region's inhabitants and authorities; overcapacity of public transportation in order to meet peak hour demand is very costly.
4. Time and money spent on commuting in the area is a huge burden for the national economy, with a social cost amounting to nearly 10 per cent of France's GNP.

Favourable conditions for telework development

1. High activity rate, and especially in the tertiary sector, where 69 per cent of employment is to be found.
2. Skills, level of education and increase in the number of graduate workers; professionals and managers are twice as many as in the rest of the country; 67 per cent of computer engineers live in the area.
3. Telecommunication equipment is the most dense in the country; the region is at the heart of all its data transmission links.
4. Aspirations for bringing closer family life and professional life are quite widespread in the region.
5. Isolation can be more easily alleviated where population and housing densities are the highest.

Key variables and decisions

1. Telework will concern mostly two workers' profiles:
   - creative tasks, performed at home, with various professional status,
   - repetitive tasks performed in neighbourhood work centres or decentralized agencies by employees of large corporations or administrations.

2. The working environment, and especially the structures of corporations and administrations (size of units, job definition, evolution of subcontracting, level of education) will decide which of these types of jobs (or if both) will develop.
3. Availability of cheap, powerful, portable micro-computers is the only technological variable to play a decisive role.
4. Urban policies, zoning laws, etc. will be critical for the economic viability of telework at home in cities which draw a large part of their resources from employment taxes.
5. Telecommunications tariff, especially for local calls, might be the major obstacle, as long as they cannot economically compete with direct costs of transportation.

Three remarks to conclude:

1. It is not enough to prove that telework is the best way to get work done. Productivity, competitiveness and other economic considerations are not the only factors to be taken into account. Organizations, public or private, have many other goals, sometimes conflicting, among which the conservation of existing power relationships and symbols of hierarchies are not negligible [24].

2. In a highly centralized country such as France, it is difficult to consider that telework on a large scale, with all its decentralizing effects, may happen without any interference by local or national authorities.

3. Autonomy is a rising value today, and the major individual drive towards telework. But it cannot be enacted by law. Education is the strongest way to develop it, but it is a long-term process which may bear its fruits in the next generation only.

REFERENCES

[1] Futuribles, Tendances d'évolution socio-culturelles dans les pays de l'OCDE à l'horizon 2005 (Paris, 1985)

[2] IPSOS, Scénarios pour demain (Paris, Feb. 1984)

[3] Beer, A. de, Blanc, G., L'avenir du travail à distance dans la région Ile-de-France, une étude prospective (Futuribles, Paris, 1985)

[4] Manning, R.A., Alternate work site programs (Control Data Corporation, Nov. 1984)

[5] Mertes, L.H., Doing your office over electronically, (Harvard Business Review, March-Apr. 1981)

[6] Webb, M., Equal opportunities for women in employment (Employment Gazette, Aug. 1983)

[7] BURSENS, L'ordinateur au bureau? ... ou à la maison (Agence d'Arc Inc., Montreal, 1985)

[8] Boscher, F., Duflos, C., Lebart, L., Conditions de vie et aspiration des Français, premiers résultats de la sixième phase d'enquête, Consommation - Revue de socio-économie, 2

(CREDOC, Paris, April–June 1984). See also Le Monde, Feb. 19, 1987, p. 28

[9] Commissariat Général au Plan, Comment vivrons-nous demain? (La Documentation Française, Paris, 1983)

[10] Halary, C., Ordinateur, travail et domicile (Ed. Saint-Martin, Montreal, 1984). See also: Adriana, F., Computers at home: new spatial needs? – a case study (Department of Architecture, MIT, Cambridge, 1982)

[11] Gassot, Y., Prospective de l'intégration des terminaux dans l'habitat (IDATE, Montpellier, 1983)

[12] Mercier, P., Plessard, F., Scardigli, V., La société digitale (Seuil, Paris, 1984)

[13] Préel, B., L'éternel retour à la maison, Futuribles no. 102 (Paris, Sept. 1986)

[14] Prospective du travail à distance en Ile-de-France, op. cit.

[15] see Kraemer, K., Telecommunications/transportation substitution and energy conservation, Telecommunications policy (March and June 1982)

[16] For a discussion of this point see Claisse, G., Transports ou télécommunications, les ambiguités de l'ubiquité (Presses Universitaires de Lyon, Lyon, 1983)

[17] Hall, E.T., The hidden dimension (Doubleday, New York, 1986), The dance of life (Doubleday, 1983)

[18] Forster, E.M., The machine stops, in: The eternal moment (Harcourt, Brace & Co.)

[19] BCEOM, Etude de substitution transports-télécommunications (DATAR, Paris, 1970)

[20] Mendras, H., Le social entraine-t-il désormais l'économique? Futuribles no. 105 (Paris, Dec. 1986)

[21] Toffler, A., De '1984' au bureau de l'an 2000 (Le Monde Informatique, Jan. 9, 1984)

[22] Prospective du travail à distance en Ile-de-France, op. cit.

[23] Renfro, W.L., Second thoughts on moving the office home (The Futurist, June 1982)

[24] Mintzberg, H., Power in and around organizations (Prentice-Hall, 1983)

TRENDS OF DECENTRALIZATION OF WHITE COLLAR ACTIVITIES BY MEANS OF INFORMATION AND COMMUNICATION TECHNOLOGIES

Helmut Drüke, Günter Feuerstein, Rolf Kreibich

Institut für Zukunftsstudien und Technologiebewertung,
IZT, D-1000 Berlin 30, Stauffenbergstraße 11-13

1. INTRODUCTION

The following explanations are based on provisional results of a study carried out by the "Institut für Zukunftsstudien und Technologiebewertung" Berlin (IZT Berlin) (Institute for Future Studies and Technology Assessment), commissioned by the "Rationalisierungs-Kuratorium der Deutschen Wirtschaft e.V." (RKW) ("Rationalization Board of the German Economy incorp.") on the subject "Büroarbeit im Wandel - Dezentralisierung von Angestelltentätigkeiten mit Hilfe von Informations- und Kommunikationstechnologien" (Changes in Office Work - Dezentralization of White Collar Activities by means of Information- and Communication Technologies) (1). As public discussion in the Federal Republic of Germany on telework, especially on some special forms such as tele-homework or work in neighbourhood- and satellite offices, is more characterized by remarkable repetition of well-known statements rather than by new empirical facts, it was time to examine present trends in the different branches of economy more thoroughly, in quality as well as in quantity. As a matter of fact, the amount of scientific publications on this subjects is also strongly disproportioned with respect to its empiric content, so the bonmot afloat among experts, that the number of these publications is more numerus than the tele-homework jobs, characterizes the reality of research to the point.

Nevertheless, there are good reasons why much attention is paid to this subject. They do not only refer to tele-scenarios of different futurologists as Alvin Toffler (2), John Naisbitt (3), or Nike Macrae (4), or to the prognoses by US-American think-factories as the M.I.T. (5), the Institute for the Future (6), or the Data Results, Dallas (7), which all are announcing an abundant development of the tele-homework jobs. With no doubt, studies which are forecasting for the next 15 years, that is to say up to the year 2000, that 40% of all US-citizens are going to telework at home, have to be taken seriously and treated as revolutionary, even if only 50% of the forecasted level were to come true. So if "only" 20% of all citizens in an industrial country were occupied by tele-homework, we would have a fundamentally changed structure in work, employment and companies. Numerous technical information- and communication patterns would underlie this structure as well as new models of job organization, changed working structures, new juridical solutions for labour, other conditions for qualification, other functions and tasks characteristics, modified social relationships as well as varied wage structures, possibilities of participation, labour protection regulations or representational models for the concerned people (8).

Consequently, it is not astonishing that predictions of abundant tele-homework caused violent debates and speculations not only amongst the "social partners", especially the Trade Unions, but also in the public and the field of science. The IG Metall unanimously demanded a "legal prohibition of electronic homework" in a first resolution on their 14th "Trade-Union Day" in 1983 (9). On the other hand, not only employers but also numerous alternative groups and avantgarde labour scientists are optimistic and highly rate the possibilities of a new formation of the world of labour and leisure time and way of living by means of tele-homework.

At this point it is suitable to anticipate the main result of our study: the actual developments are extremely modest, at least for the Federal Republic of Germany with respect to the quantitative development of tele-homework. Nevertheless, in the different branches of economy, there are extremely different developments, where telework plays an important part and probably will do so increasingly. This development is associated with large scale of variation and decentralization- but also new centralization- (or recentralization-)trends in work organization and management structure (10).

We come to the conclusion that in future the subject telework, within the scope of new firm- and employment models, which will be described succeedingly – and moreover in numerous other organization forms, nowadays still unknown, will be of great importance and will definitively change the economic and employment system. With regard to this, there are a number of specific reasons, which shall for the moment only be stated cursorily:

1. Principally there is a high potential of work functions or tasks in the industrial work and employment pattern, which can be transferred to other areas; here we follow the quantitative estimates of American "think factories", which state potentials between 40 and 70% of all working activities.

2. The technical development of communication networks and systems and their cost-structure will not form insurmountable obstacles for abundant introduction of telework structure at medium or long term.

3. A rationalization or cost minimizing potential linked with telework structures exists in the producing sector as well as in the service rendering sector.

4. From a business economical viewpoint and the management policy, strategies could become of special importance, which base on a minimum basic staff, through which problems caused by fluctuating orders, temporary activities (e.g. system development, programming, special engineering services, etc.) or social welfare plans in times of crisis, can be solved more easily in the company, and the influence of those involved in unpopular decisions is reduced.

5. Telework is favourable to the strong social and work organization trend of making working hours and conditions more flexible. In this case, the interests of employee and management go into the same direction to a large extent. The fact is that both aim at a common form in which their thoroughly different and to a certain degree contrary intents and concerns are realized. In future the question "whether" will be less im-

portant than "how", as to which administrative plausible technical and work organization solutions are most favourable to this seeming consonance.

6. Certain types of telework make it possible to meet better and faster different corporate demands, which are gaining more and more importance for the service rendering society, such as proximity to the market and to the client, consultation, servicing, maintenance, control, innovation promotion.

7. As one has to proceed from a long lasting negative employment situation in all industrial societies, telework jobs, if they seem to be useful from an economical or managerial viewpoint, will be accepted, even if they do not create an optimal employment situation for employees.

2. EMPIRICAL EXAMINATION

The situation of telework can be characterized as extremly complex, especially if not only the extreme cases "tele-homework" and "work in neighbourhood offices" are examined, but the whole spectrum of decentralization by means of information- and communication technologies. But as the IZT-study aims to elucidate all the development trends, a differentiated methodical start had to be chosen. As to quantitative and qualitative results, we strived for representativity for the German economy. It was clear from the beginning, that it was not a question of a numerical mechanism, which seems to locate every telejob exactly, but a question secure variations, in which developments elapse in their system multiplicity. So the identification of performed and intended decentralization had priority over the ascertainment of the precise number of teleworkers.

Methodical Proceeding

The methodical proceeding, along with analysis of international literature and reports on telework, covers the following field approaches:

- a companies inquiry of around 3ooo corporations from 22 economic branches,

- intensive interviews of 18 experts from different specific fields (industrial economics; technical science; labour market; labour law; family and leisure time; ecology from different institutions (state, economy, trade unions, politics, research)), different problematical situations (infrastructure installations, producers, consumers) and constellations of those affected,

- case-studies in 2o companies, selected from various economical branches, aimed at obtaining detailed knowledge on different typical decentralization models by means of information- and communication technologies in the white collar field with regard to their technical, social, economical, legal and managing-strategical character.

The total investigation is designed as a panel study. The companies inquiry as well as the case-studies will be carried out again after 1 1/2 years. The following explanations are based on the result obtained from the first phase and from the interviews already carried out with experts.

Notes on the Method

The multistage methodical proceeding has been also advisable since neither the companies nor the affected employees nor their representatives were likely to have much interest in giving information on telework development. An important part is played here by the fact that in the companies is not much will to talk about technical and organizational systems which are still being planned, developed and tested. There is also great uncertainty with regard to internal welfare consequences, on expense structures, on qualification requests, legal conditions, etc.

On the employers side, open abd latent uncertainty and anxiety with regard to the employment situation dominate. Naturally the safeguarding of jobs is most important, so that all information which might actually or apparently endanger them, are given only very unwillingly.

In order to achieve a participation in the inquiry of as many companies as possible despite these difficulties. we reduced the questionnaire to questions about already performed or intented decentralization by means of new information- and communication technologies. We also asked companies for detailed information on employed technical systems or telecommunication services, which are used for decentralization, as well as on the organizational form of decentralization by means of information- and communication technologies.

The selection of the economic branches should include the most important branches of the production industry and services sector. Besides the branch-specific concentration of activities suitable for decentralization of white collar activities by means of information- and communication technology or those where these forms of decentralization are being tested or practised should be reflected.

Furthermore, when selecting branches, it is also taken into account that **different task forms with different qualitative and quantitative demands** can be included in the investigation. At last economic branches, which have a **key position in development of electronic telework**, are favourably taken into account. According to these criteria 22 branches have been selected for the branch inquiry.

3. RESULTS

3.1 Comprehensive survey

The following chart is totalling the result of the company inquiry. About 3000 from 22 economic branches were questioned by us on already performed or intended decentralization of white collar activities by means of new information- and communication technologies. The exact number cannot be stated, as the questions were directly sent from central professional associations to four branches: wholesale and foreign trade, mercantile agents, plastic material industry and management consultants.

The IZT itself has sent away 2268 questionnaires. 186 (=22,9 %) of the 812 companies which sent back the filled-out questionnaire (that amounts to a return quote of 35,8 %), stated they already had carried out or are planning a decentralization. More than half of these, that is to say 126 companies (=15,5 %) stated to have carried out a decentralization of white collar activities by means of information- and communication technologies. 81 companies (10 %) stated they would take such measures at medium or long term range.

3.2 Branch survey

Positive answers came most frequently from those branches, which seemed especially suited for decentralizations by means of new information- and communication technologies after preparatory works. To these especially belong:

- media (news agencies, broadcasting companies, newspaper publications)
- insurance companies
  and in the scope of the anticipated uncertainty
- trade (wholesale and foreign trade, mercantile agents)

These branches are already decentrally oriented according to their activity field character. In other words, foreign services activities or a mixture of office and mobile work are important defining features of these branches, before even using new information- and communication technologies.

The new information- and communication technologies improve mobile work and communication amongst foreign services staff and central office or central office and customers (trade). Another group with a high proportion of performed decentralizations of white collar activities by means of new information- and communication technologies has to be placed in the column "Branches with branch-offices/ agencies system":
- saving-banks
- local health insurances
- air-transportation companies

Branches which have displaced activities, but are still working under off-line-management and so create an important potential, belong to another column. These include:

- printing industry
- insurance agencies
- newspaper publishers
- news agencies
- broadcasting stations

The printing industry and insurance agencies for the service rendering sector play a special pace-maker part here and are of great importance.

For several branches decentralization - at least up to now - is less relevant:
- automobile industry
- banks
- chemical industry
- iron and steel industry
- electrical industry

- aviation industry
- mechanical engineering

These branches with no doubt belong to the basic industries of the German economy and mainly apply to the production industry. So one could draw the cautious conclusion that decentralization of white collar activities mainly takes place in the service rendering sector and still plays a secondary part in the production goods sector and investment industry. The reasons for a relative hesitation of the ventures of these branches have been detected with more preciseness in model studies.

Some fields of the production industry and the state were especially interested in the decentralization of white collar activities by means of new information- and communication technologies, which was shown by statements of corporations and authorities as to relevant planning. Mechanical engineering has been especially outstanding. But corresponding planning is also reported from the electronical and chemical industry. A similar interest was stated by authorities, where planning with respect to decentralization of activities is characterized as being more or less thriving. Here also, model studies were able to confirm the provisional result.

4. DECENTRALIZATION FORMS

4.1 Requirements and trends

The results of the company inquiry and the model studies with respect to conditions and trends for telework correspond with estimations of experts questioned by us.

Accordingly, it can be stated that relevant flexibilization and decentralization potentials can be mobilized with new information- and communication technologies, when business administrative, work organizational or company strategical reasons are obvious. Main conditional coherences therefore seem to be:

1. Promotion of productivity and competitiveness through mobilization of rationalization, flexibilization and adaption potentials.

2. Improvement of the reacting capability of the whole corporation toward external structural changes: market adaptation, attendance of customers, exploitation of regional characteristics of the labour market, marketing chances, changes in technological and ecological conditions and others.

3. Adaptation of company structures to internal requirements such as changes in the production and services organization, technological outfit and structures, co-workers requirements, and others.

Moreover, it is obvious that these mixed forms of telework are mainly adaptable for activities, which already had a certain degree of variability before applying modern telecommunication technologies. Insofar it is mainly a matter of systematical support, strengthening and expanding of already existing structures and trends. The obtained experience becomes interesting, when it has to be transferred to less qualified activities

Results of the Corporate Inquiry

Results of the corporate inquiry

Branch	Number of valid questionnaires	Return absolute	Return in %	positive absolute	positive in %	performed decentralization 5 absolute	performed decentralization 5 in %6	planned decentralization 5 absolut	planned decentralization 5 in %6
Automobile industry	77	51	66,2	7	13,7	4	7,8	3	5,9
Banks	47	22	46,1	3	13,6	1	4,5	2	9,1
Chemical industry	151	46	30,5	9	19,6	4	8,7	5	10,9
Printing industry	271	40	14,8	13	32,5	10	25,0	3	7,5
Iron and steel industry	119	32	25,9	5	15,6	3	9,4	3	9,4
Energy	148	46	31,1	3	6,5	3	6,5	2	4,3
Elektronics	167	55	32,9	9	16,4	6	10,9	5	9,1
Foreign and wholesaletrade 1	?	(6)	—	(3)	(50,0)	(1)	(16,7)	(2)	(33,3)
Mercantile agents	(620)	(27)	(4,4)	(14)	(51,9)	(5)	(18,5)	(11)	(40,7)
Plastics industry	100	34	34,0	2	5,9	—	0	2	5,9
Air line companies	95	13	13,5	3	23,1	2	15,4	1	7,7
Aviation industry	81	31	38,3	5	16,1	4	12,9	1	3,2
Mechanical engineering	172	49	28,7	8	16,3	2	4,1	7	14,3
News agencies	7	2	28,6	2	100	2	100	—	—
Public services	103	64	61,2	10	15,5	7	10,9	9	14,1
Local health insurances	90	37	41,1	7	18,9	5	13,5	2	5,4
Tourist agencies	80	11	13,8	—	0	—	0	—	0
Radio stations	14	9	64,3	3	33,3	3	33,3	2	22,2
Savings-banks	118	43	36,4	12	27,9	12	27,9	—	0
Company advisers 3	(290)								
Insurance companies	301	169	52,8	39	24,2	24	15,0	19	11,0
Newspaper	126	67	53,2	46	68,7	34	50,7	15	22,4
Total 4	2268 (3178)	815 (845)	35,8	186 (203)	22,9	125 (132)	15,5	81 (94)	10,0

1 The exact number of corporations questioned by the "Federal Union of the Wholesale and Foreign Trade" could not be identified
2 620 corporation were questioned by the "Central Association of German Mercantile agents and Trade Organizations"
3 All member corporations (about 290) were questioned by the "Bulletin of the Organization"
4 Sum of all questioned corporations
5 Incl. double naming as several corporations have performed as well as are planning decentralization
6 Referring to positive returns.

and less privileged positions. Not only because of the quantitative significance, but also because of increasing work organizational, ergonomical, labour juridical and social problems. The process of decentralization is even less spectacular in all those fields, where work was traditionally decentral (example: insurance economy, trade, etc.).

Through the application of new information- and communication technologies, spatial decentralization of activities is strengthened and generally this does not lead to less control in the head office or a significant expansion of autonomous discretion at the decentral premise. But in most cases it leads to substantial changes in the relationships between the involved parties, for example between central office and branch offices, between firms and their suppliers and finally between the particular employees amongst each other. The most important quantitative and qualitative potentials of decentralization can be found here.

On the whole we are facing a continously changing work and venture structure, where manifold decentralization trends and also decentralization processes originated by telework can be noticed, but where classical forms of telework at home and neighbourhood and satellite offfices only form a small part of the total spectrum. Consequently, the telework-typology used up to now does not meet the manifoldness and importance of different forms of telework-trends. In consequence, for the moment we renounce attempting to create new definitions and will only differenciate between decentralization forms within and outside the limits of corporations. The following examples are to make plain the tendencies.

4.2 Telework within a company

Here, the organization of work generally changes aggravatingly, but the company remains the same as before displacement.

In all these cases, activities or functions are shifted, which also change in part, but this process is not necessarily joined with the displacement of persons.

Examples

- within an existing and spatially defined company work is integrated and heterogenous activities are combined in one hand (e.g. case specific or cliental referred comprehensive revision in case revision damage and credit winding-up);

- activities/functions carried out centrally in the past are transferred to already existing branch-offices;

- activities being transferred to newly founded branch-offices next to clients and/or near to residences of labour forces (satellite offices);

- activities changing from internal to external service (e.g. as mobile telework including computer supported preparation and, given the case, process revision); the most salient examples are insurance agents and journalists as mobile teleworkers and mobile telebusses from local health insurances as "flying branch-offices";

- Indoor-assistants occasionally take over the local attendance of customers (such as in the form of "mixed-jobs-employment")

- Permanently employed assistants with more skilled qualifications (managers, developing engineers, etc.) obtain along with their usual place of employment (second ofice or place of employment at home, neighbourhood office)

4.3 Telework outside the limits of a company

In these cases the division of work is changed socially by regroupings among the companies.

This development belongs to one of the most spectacular trends in telework, as here can be seen first approaches for a new diversification of the company structure.

Examples

- work functions are not carried out within the company itself, but through independent companies (e.g. office service administra-tions, neighbourhood offices. software houses), which function as subcontractors;

- supraoperational installed services organisations (e.g. the automatic reservation system START of West-German tourist agencies) take over activities, which up to now were carried out by individual companies. A decentralization of individual companies is from the associated organisation's point of view a centralization;

- work functions are displaced to persons who carry out their tasks outside the company premises and do not have regular work terms, such as telework (at home) in the work contract, franchise contract, as "person simular to employee", in the home employment contract, in the loan employment contract;

- the customers (company, free lance, private persons) take over work in self-service, which was performed by personnel before (tele self-service e.g. with teletext).

Finally, is has to be pointed out that such forms also exist on an international level, which provides the trend with extraordinary dynamics and leads to other numerous problems in consequence.

5. DEVELOPING TENDENCIES

5.1 Activities which can be decentralized

In discussions concerning development trends of telework, various general criteria such as economy, work organization, new data-processing and transfer techniques are called cooperator's motivation, with which extent and transposing dynamics of telework potentials are substantiated. Heilmann (11) takes pattern from "job characteristics" by M. Olson in the discussion on decentralization potentials, which he describes "as pragmatic characteristics of suited activities for telework":

- a small demand of resources
- an individual work speed
- a clearly defined result
- an important mental concentration
- time-schedule and work quantities
- a small constraint for communication

It is noticeable that contrary to earlier discussions in scientifical publications and in the public, the business economical-pragmatical aspects respective of introducing telework, i.e. the forms of decentralization in companies, are of much greater significance. Amongst others, the following belong to these aspects:

- presumable rationalization effects, competitive trade benefits, cost benefits
- improved marketing chances, greater marketing proximity, proximity to customers
- more effective work organization through improved communication relations, working progress etc.
- improvement of working results through improved concentration of the people involved, improved adapted qualifications, better motivation and greater flexibility towards the challenged requirements.
- the capability to routinize and to standardize activities
- solution of allocation- and transport problems

Only from this schedule it is obtainable that the trends with regard to classical division of qualification (superior and inferior working aktivities) by no means show towards the same direction. On the contrary, one of the most important results of the study is that telework can apply to qualified activities as well as to simple routine activities, for example activities that can be routinized and standardized. In other words,the qualification level of activities and functions is not a suitable criterium for the probability of telework. Other characteristics have to be taken into account, which accelerate or hamper displacement of white collar activities by means of new information- and communication technologies in the precise case. Aspects which refer to internal work organization and motivation of the involved also form part of these, such as

- the desire to flexibilize work
- status improvement by using technical systems
- requirements of direct consultancy
- the type of control and security problems
- social contacts and qualification possibilities
- data protection and data security problems

Without striving for completeness, the following activities can be classified as being especially suited for decentralization trends through telework:

- programming activities
- mobile activities (mercantile agents, insurance agents, journalists and alike)
- designing and constructing
- interpreting and translation
- the work of judges, lawyers and university teachers
- the advising task of staff in companies
- marketing

- servicing
- text processing, data processsing
- manifold revising activities

Revision is an especially controversial case. Not every type of revision can be presumed as being suitable for telework. The above mentioned criteria very much differentiate the wide range of "revision", so that, in principle, definite statements are only possible for certain forms of revising activities.

Activities such as text processing and data processing, which in the near future will be carried out partly or completely by machines, can be classified as being only temporarily executable in telework.

5.2 Accelerating and hampering factors/combinations of factors

The above discussion of characteristics of working activities and functions, which form favourable or unfavourable conditions for decentralization trends by means of new information- and communication technologies, caused several accelerating and hampering factors. Nevertheless, these are not yet sufficient conditions for the actual introduction of decentralization models for white collar activities. Therefore we are going to state other relevant factors or combinations of factors, which accelerate or hamper telework. Again are to be emphasized business administrative considerations on economy, work organization, competitiveness accelerating factors for motivating co-workers.

The factors market proximity, attendance of the customer, acquisition, servicing, quality control, personnel planning at medium and long terms are becoming more and more important for ventures. Here it only can be examined and decided in detail if and which way the corresponding factor or combination of factors has an accelerating or hampering effect on decentralization forms of telework. It can be seen that only these few factors can already lead to a highly complex dicisionary structure.

Additionally there have to be stated some of the conditions, which are of special importance due to a special state of interest of the involved and due to general social-political conditions.

- The seeming harmony of interests of employers and employees:
  The fact that there are a great number of constellations, in which interests of employers and employees with regard to introduction of decentralization forms by means of telework are apparently conform, is an extremly strong and accelerating momentum at the present state of technical sciences but also generally for future development.

  In the case of less qualified activities these are often based on an emergency, either to find a job in structurally weak areas or - especially in the case of women - being able to provide for the family or children. Goldmann and Richter are stating that for some women the possibility to telework under present individual conditions seems favourable by no way of substitution.

In the case of more qualified activities the actual and the supposed gain concerning independence, autonomy, etc. plays an important part in the decision of starting to telework, either part- or fulltime work.

All types of mixed work become especially important for all qualified areas, because this way can be avoided or reduced considerably most of the negative factors such as isolation, internal administrative career losses, etc. .

- The situation on the labour market
  There is no question about the fact that the total economical situation and especially the situation on the labour market - especially branch-specific - accelerates or hampers development substantially. But also here, there is no definite interdependence. On one hand, in the case of substantial unemployment, employees - especially in structurally weak areas˝- will accept telework jobs even when they do not correspond to their ideas; in the same way, corporations will try to make extensive use of all rationalization, cost and competitive advantages- and above all reducing staff - by means of telework.

  On the other hand it is likely, that risky experiments and pilot projects are avoided and the purchase and application of a not yet developed and expensive technical science is discarded. Employees, especially those of higher qualifications, who seek more independance, will link more to traditional work forms for reasons of job safeguarding internal administrative qualification and career.

- The trade union policy
  The foregoing elaborations have proved that companies are facing complex situations of decision. But this situation is even more complicated for trade unions. There is no doubt, that the nature of the entire problem is in the first place, that trade unions nowadays have no uniform and convincing policy in this field. Problems in the particular economical branches are very different; numerous contrary economical, business administrative, qualification-specifical and motivational factors give a complete picture difficult to overlook. Nevertheless, trade unions will be faced more and more with the problem and a lot will depend on what policy the trade unions adopt.

- The development of technique
  The state of techniques, especially the underdeveloped state in public networks and services, is seen as important hampering factor for further decentralization of white collar activities by means of new information- and communication technologies. But there is no uniformity at all as to this subject. Who considers the aspects of economy, of work organization, of qualification and control in connection with decentralized work structures as the decisive aspects, will rate less important the effects of technological changes subsequently to development of the telecommunication networks than those who consider the present efficiency of telecommunication networks as being deficient, because for the latter these are the decisive requirements for a masssive introduction of telework structures under conditions which favour this introduction of techniques and cost structures.

But there are quite different opinions on the decisive questions for which purposes (in social and private communication), to what conditions (responsibilities, what type of network, relative openness or larger standardization) and to what prize (financially as well as communication-ecologically).

According to the present findings, we tend towards the following interpretation: Decisions on rate, type and conditions for development of public networks have essential effects on rate and forms of decentralization of white collar activities.

A generally highly developed network is favourable for services, that is to say, all those which have a widely branched net of foreign services co-workers as well as those who aspire to a displacement of activities of the working population on private consumers. So mobile telework as well as electronic self-service here will surely be encouraged, whereas the form of productive closed user groups with their own service characteristics (e.g. in big production corporations) is more likely to be carried out through inhouse-networks. Generally, we think technological development is so dynamic, that on medium terms it is more likely that new possibilities are created through them than barriers are built up.

- Data protection and data security
  Due to our own surveys, we can generally state, that in all cases of decentralization by means of new information- and communication technologies, where data transfer to third persons take place or networks and systems with access for the public are being claimed, problems on data protection and security occur. As this is the case in almost all forms of decentralization, these models of decentralization have to be treated carefully with regard to this.

Data protection commissioners see the problems as so aggravating, that for different forms of decentralization the fundamental question as to legalness of data transfers through open networks beyond pure data processing is doubted. But without doubt, there still exist essential normative deficits for almost the whole spectrum of decentralization. (12)

The problems of data security against a manifold spectrum of possible and already recognizeable leaks in the systems and especially in application by involved persons, can be very difficult for individual systems and play an important part as hampering factor for displacement of white collar activities.

6. RESUME

6.1 Need for regulations

Principally, regulations can be made on four levels:

- on an economical or social level
- on a collective juridical level with operational agreements
- on the labour protective field
- on the individual juridical field, that is to say on a level of direct employer-employee-relation.

The results of our study suggest that the priority of measures corresponds to the order of these four levels. Principally, there is a need for regulation on all four levels. In the meantime, there are numerous, partly extremly contradictive suggestions on all four levels. Succeedingly, only regulations which seem especially important on ground of the study, will be emphasized.

Solutions to the problems for decentralization of white collar activities by means of new information- and communication technologies naturally depend decisively on the requirements as listed in 5.2. on the totally political level. It should be clear to all involved and responsible persons in the political and "Sozialpartner"-field that economic, employment and technological as well as juridical conditions decide on principle, if soft solutions or hard conflicts take place in the social and political structure of the Federal Republic. Radical changes in labour and company structures associated with a massive intriduction of telework can only take place without greater frictions, if

- entire economy conditions are good and are not based on the costs of minority groups, especially of unemployed men, women, less qualified, disabled or individual professional groups and
- development and participation of techniques are oriented on essential needs in the order of reinforcing public welfare and individual autonomy or general improvement of life and labour quality.

This implies political need of regulations for

- qualitative selective economic growth
- general offensives for full employment and qualification
- social engineering development and application
- special requalification concepts for potentially involved employee groups

Regulations on the collective juridical level are of special importance. Science could carry out an important preparatory work here, a stock-taking of claims, especially on industrial agreement regulations and administrative agreements scheme. As we are dealing with a large field, an individual regulation catalogue has to be set up for each decentralization form only some examples of problem areas shall be stated here:

- qualifications and strain change
- loss of jobs in indoor service
- jobs of women are especially endangered ("defeminisation of the office")
- hardware and software of decentralized jobs cause aggravating problems
- administrative personnel is disqualified during the automatization process
- electronic clientel self-service endangers jobs

Without going into details, the practice up to now proves that collective-juridical regulations with regard to rationalization protection agreements, mostly have a hampering, rarely a developing effect. This relative protection is only valid for existing labour contracts. Therefore it does not hamper establishment of telework in undesired ways of independence. Even regulations which literally are to be understood as protection (as e.g. the institution "person, similar to employee"), prove as being ineffective in practice. The few wage agreements according to § 12a "Tarifvertragsgesetz" (wage agreement law) can be evaded without consequences. The main point of a new wage policy should be the closer link of quantitative and qualitative wage policy, in order to cope with labour and qualification structures, which have been changed through application of technology in a standardized concept.

Catchpoints:

- requalification, that is revalorization of certain activities without qualification losses for certain employment groups (e.g. associating assistant and revising functions);
- humanitary formation of working and living conditions, that is, avoiding overtaxing and undertaxing, example: "Lohnrahmentarifvertrag" (wage scale agreement) of the IG Metall in Nordwürttemberg/Nordbaden;
- humanization of work through labour-oriented formation of work and techniques.

It is evident, that such measures are better for collective decentralization forms such as e.g. neighbourhood and satellite offices as for telework at home. They also meet the wish for a meaningful association of private and work life. In these cases a labour unionist representation of interests - a central social problem in the case of decentralization - should also work. As there is a large discussion in the field of labour protection and about individual regulations on the level of direct relations between employers and employees, we only refer to relevant literature at this point.

During the study we obtained numerous interesting aspects. Thus, we commented on problems concerning the companies inquiry and the case studies.

The question of applying administrative and superadministrative co-determination concerns many aspects. Here one can say:

- principally, an expansion of super-administrative co-determination rights should be achieved so that influence on part of the involved employees on a social-contractual introduction of telework and decentralization measures is essentially strengthened;
- expansion of co-determination in the "Betriebsverfassungsgesetz"(works industrial-relations-law) could also help to avoid, that an essential part of the employees is not forced into second-class working conditions. Nevertheless, one has to be sceptical on the application of these regulations, as the organizations of employees (factory councils and shop stewards) have not even managed to exploit the "Betriebsverfassungsgesetz."

Further on, one has to be sceptical with regard to the possibilities of new legal regulations for extending codetermination. For example, the unlimited codetermination of the personnel council "in the case of placing or privatization of works or activities, which were responsibility of the employed in the office up to now" were annulled by the law-court of Hessen.

6.2 Summarizing theses

1. Decentralization of white collar activities by means of new information- and communication technologies is an extremely complex problem, which is going to influence and change labour, employment and corporate structures qualitatively and quantitatively, at present and in the future.

2. In the future there will be manifold forms of decentralization by means of new information- and communication technologies. Classical tele-homework will only form a small part.

3. Not only less qualified activities and labour functions are suited for dislocations, but above all qualified employed, for which a separation from the central office is sometimes necessary and useful, will make more and more use of decentral telework.

4. Decentralization of white collar activities is especially plausible, where along with business administrative advantages (rationalization, cost lowering, better work organization), company-strategic aims (competitive advantage through more market proximity, cliental proximity, improved service, product flexibility, innovation, servicing) can be better achieved.

5. Decentralization is especially favoured, where there are apparently equal interests of employers and employees, that is, where business administrative advantages and company-strategical aims coincide with the employees' whishes for more temporal flexibility and autonomous formation and decisionary scope without causing fears or insecurities concerning jobs.

6. Experiences so far seem to prove, that the future of decentralization of white collar activities by means of new information- and communication technologies lays in the "multi-objective jobs", to which satellite offices also have to be added. Potential linking of employees with the central office reduces isolation problems, qualification and career losses as well as fears to be disqualified as second-class employee.

7. A highly interesting development trend is noticeable with regard to a new company structure. Increasingly, large companies tend to install their delivery and service structure to "independent" sub-corporations. These will then only be linked with central offices by inhouse- or public networks. Along with numerous advantages with respect to product, service and transport shaping, this structure reduces numerous staff-economic problems.

8. There are numerous accelerating and hampering factors with respect to further decentralization of white collar activities by means of information- and communication technologies, of which the most important determinated by whole-economical and political basic conditions.

The expansion of public telecommunication networks and the improvement of the resulting appliances is often characterized as "conditio sine qua non", but actually has only an accelerating or hampering effect and only a marginal influence on some forms of decentralization.

9. There is a need for regulations on all levels. Most important is the total economical or political level which not only decides on decentralization together with new labour-, employment- and company- structure, but also on the power balance between the "Sozialpartner" (employers and unions) and on the future of trade unions.

There is also a need for regulations on the collective-juridical level, in the field of labour protection and in the individual juridical field as well ays with regard to data protection and data security problems. The latter does not imply principally barriers for further decentralization of white collar activities.

7. REFERENCES

(1) H. Drüke, G. Feuerstein, R. Kreibich, Büroarbeit im Wandel - Tendenzen der Dezentralisierung mit Hilfe neuer Informations- und Kommunikationstechnologien, RKW-Schriftenreihe Mensch und Technik, Eschborn 1986.
(2) A. Toffler, Die Zukunftschance (Die dritte Welle), München 1980
(3) J. Naisbitt, Megatrends, New York 1982 (German edition, Bayreuth 1984)
(4) N. Macrae, The 2024 Report - A concise history of the Future 1974-2024, London 1984
(5) M.I.T., Teletext and Videotext in the United States, New York 1982
(6) quoted: Job 2000, Arbeitsplatz zu Hause, in: Psychologie heute, September 1982, p. 7
(7) Data Results, Reprints No. 11, Dallas 1983
(8) ref. to reference no. 1; additionally: H. Drüke, G. Feuerstein, R. Kreibich, Dezentralisierung von Angestelltentätigkeiten mit Hilfe neuer Informations- und Telekommunikationstechnologien
   - Expertengespräche zur Telearbeit und Tele-Heimarbeit, Paper: Berlin, December 1986, in print
   - Modellstudien zur Telearbeit und Tele-Heimarbeit, Paper
(9) IG Metall, 14. ordinary Congress of the Trade Union, Resolution No 9, Neue Informations- und Kommunikationstechniken, Pt 6, October 1983
(10) ref. to reference no. 1, p. 37ff
(11) W. Heilmann, Elektronische Heimarbeit - oder die Politisierung der Heimarbeit, in: Die Betriebswirtschaft Nr. 1, 1985
(12) particularly H. Garstka, Collaborator of the Berlin Commissioner of Data Protection, Interview, ref. to reference no 8
(13) the edition of the case studies will contain more detailed particulars

5

TELEWORK FROM A EUROPEAN VIEWPOINT

TELEWORK IN THE EUROPEAN COMMUNITY: PROBLEMS AND POTENTIAL

by Eberhard KÖHLER, Rosalyn MORAN, and Jean TANSEY

Eberhard Köhler is Research Manager in the European Foundation for the Improvement of Living and Working Conditions, Loughlinstown House, Shankill, Co. Dublin, Ireland. Rosalyn Moran and Jean Tansey are associated with the Irish Foundation for Human Development, Garden Hill, 1 James's Street, Dublin 8, Ireland.*

Abstract: This paper reports on the main findings of telework research projects which were sponsored by the European Foundation for the Improvement of Living and Working Conditions between 1982 and 1986. Questions of the real diffusion of telework as well as questions of inhibiting or favourable factors influencing its future development are discussed from a European Community perspective.

1. INTRODUCTION: THE EUROPEAN FOUNDATION FOR THE IMPROVEMENT OF LIVING AND WORKING CONDITIONS

The European Foundation for the Improvement of Living and Working Conditions is an autonomous Community body established by a Regulation of the Council of Ministers of the European Communities which came into force on 26th May 1975.

The Foundation was created in recognition of the fact that problems associated with improving living and working conditions were growing more numerous and complex, that Community action to resolve them should be built on scientifically-based information, and that the social partners should be associated with such action.

The creation of the Foundation in the mid-1970s was therefore a complementary development to national efforts in the Member states of the European Community, where programmes or agencies for the improvement and humanization of working conditions were promulgated (e.g. ANACT = Agence Nationale pour l'Amélioration des Conditions de Travail in France; HdA = Humanisierung des Arbeitslebens in Germany, IACT = Institut National pour l'Amélioration des Conditions de Travail in Belgium).

The mandate of the European Foundation, however, went one step further by not being restricted to the improvement of **working conditions** alone, but also including the improvement of **living conditions.** This is reflected in its charter, where five areas of concern are enumerated:

- man at work
- the organisation of work and particularly job design
- problems peculiar to certain categories of workers
- long-term aspects of improvement of the environment
- the distribution of human activities in space and in time

---

*This paper is based on research sponsored by the European Foundation for the Improvement of Living and Working Conditions. However, the views expressed are those of the authors and do not necessarily represent an official position of the European Foundation.

Practically all these main areas of concern are touched upon by the phenomenon of telework. The relevance to the area of work is self-evident, the relevance to the environment is given by the fact that telecommuting is much less polluting to the environment than real physical commuting by car or public transport; and as far as the distribution of human activities in space and in time is concerned, telework can serve as a model for adaptability and flexibility as far as "consumption" of either of these goods is concerned. It is not surprising then, when in the framework of its second four-year programme (1980-84), which was developed along the lines of the general theme of "impact of technological development", Telework became one of the Foundation's specific research themes.

2. EUROPEAN FOUNDATION'S APPROACH TO TELEWORK

The first of the Foundation's telework projects was developed and commissioned in 1982, background and field research was then conducted in 1982/83 [1]. The objectives of the study were defined at the time within a context of many predictions based primarily on assumptions but on very little research results. Enthusiastic predictions for telework had been made everywhere. In 1971, American Telegraph & Telephone (ATT) predicted that most Americans would be working at home by 1990; two studies dating from 1974 - one Japanese and the other British - stated respectively that 65 % of jobs in the tertiary sector and 20 % of all jobs could be performed remotely. And many thought, this would actually happen imminently.

The Foundation's research guidelines therefore set the goal that this project should attempt "to ascertain the present scale and characteristics of telework in Europe and the United States, and to analyse its impact on living and working conditions and the manner and conditions of its development."

The following short summary of this study is presented here, although some of the conclusions seem already dated, whereas some of the then real experiments with telework have folded. But some of the findings in 1982/83 are of more than historical interest and show that in particular some of the social scepticism of the time has not only persisted, but many have increased since then.

In the project of 1982/83 - Telework - Impact on Living and Working Conditions, it was found that the number of applications of telework and persons concerned was very small: a few thousand people in Europe and the United States together. This corresponded to the current predictions by the experts, who had thus already moderated the forecasts of the 1970s.

In spite of their small number, these applications of telework covered a wide variety of situations. Two main types of telework promoters were identified: commercial firms, both privately and publicly owned, and the State (central government and regional and local authorities). Europe differed from the United States, particularly in the important role played by the State.

A number of underlying objectives common to most telework case studies could be ascertained; however, there were different emphases on these dimensions from country to country.

2.1. The politico-economic systems dimension

Telework was regarded as a planning tool for decentralization and revitalization of areas receiving economic assistance, for example, in France, Sweden, Italy and the Federal Republic of Germany. The main protagonist of such projects was the State, helped by the public sector economy and sometimes also by firms in the private sector (as in the Nykvarn experiment in Sweden).

2.2. The work organization and rationalization dimension

Telework was seen to improve productivity (as with insurance agents, Donnay in France, and Continental Bank, Control Data and Freight Data in the United States), to cut operating costs (DGI - the French Inland Revenue - and Actel de Cannes, in France) and allow additional pools of labour to be tapped and utilized (Legrand and the DGI in France). It was found to achieve flexibility in personnel management and cut social costs (United States).

2.3. The social dimension

Telework was regarded as a potential instrument for the improvement of working conditions; it was used, for example, by the French Telecommunications Corporation to mitigate the adverse effects of automation. It may have been a response to specific personnel management problems. On a more general level, it was regarded as a potential instrument for improving the quality of life.

The actual projects implemented and described differed from each other significantly:

- Planned, "state induced and state-sponsored" telework is the most spectacular, but also the most theoretical. Its aims are connected with politico-economic and social projects.

- "Problem-solving" oriented telework seems to be the most common form of development, the aim being rationalization of work organization.

- Unsystematic, unplanned, "creeping" introduction of telework as a small-scale decentralization project is characteristic for a large number of the cases studied in France, the United Kingdom and the United States.

As far as the persons affected were concerned, the impact of telework varied enormously from country to country and even within a single country, according to the form of remuneration chosen and the contractual relationship, location (work at home or at a telework centre), and the importance of telework in the worker's activity (work shared between firm and home, part-time telework and full-time telework).

Telework concerned specific **populations:**

- executives in the employ of firms and independent and freelance workers (executives, programmers, analysts, research workers, etc.). Telework then performed a supplementation function, as a specific form of work organization intended to rationalize and/or facilitate a pre-existing flexibility of location and working hours.

- less skilled workers, usually women: typists, enquiry operators, file maintenance staff, etc.

Most of the workers concerned in France were monthly-paid. On the other hand, in the United States and Britain piecework was not uncommon, as well as performance bonuses and the subcontracting of work to temporary agencies or associations of independent workers. Part-time telework, with less social welfare cover for the workers concerned, seemed to be most common in these two countries, probably because telework at home is much more widespread there than in France, Sweden or Italy.

The **jobs** concerned were in the fields of data processing - in particular, programmers and analysts (teleprogramming project in Germany), data acquisition, office jobs such as typing (Telecommunications Corporation in France, typing pools in Sweden, Investor Diversified Services Incorporated, Minneapolis, etc.) and file maintenance (in the insurance sector, as at Blue Cross/Blue Shield of South Carolina).

As far as an impact on general living and working conditions could be ascertained in the study of 1982/83, it was found that the degree of autonomy of low-skill jobs remained low in telework, but increased in the case of more skilled jobs, mainly through greater independence in the use of space and time.

The distance separating teleworkers from other groups in the firm did not necessarily reduce contacts with the latter but formalized them, thus enhancing their productivity. Unlike executive staff, workers regretted the depersonalization of relations resulting from their electronic communication. Specialists, who are skilled professionals, suffered less than the set-task performing staff from the reduction of information relations ; these exchanges are more functional for the former, based on a complementarity of competences, while for the latter the personal contact is a need - an essential condition for cooperation and a fundamental aspect of their work.

There was an overall change in the integration of the different parts of daily life perceived by the teleworkers. As in the case of autonomy, the arrangement of working hours depends on organizational conditions. In some situations the worker benefited from greater flexibility and saved travelling time, while in others the constraints were exacerbated, isolating the workers from their social environment, because they felt "chained" to the terminal.

Finally, in looking back to the conclusions of the Foundation's 1982/83 telework project, the forecast formulated then has a ring of surprising actuality to it today, at the turn of 1986/87. The authors concluded that it seemed unlikely that telework would expand rapidly in the immediate future. While the necessary technical infrastructure - telecommunications networks and the computer and office automation hardware - was already available or was rapidly being installed in Europe, the economic and social conditions for its development were less favourable.

After this initial project, the European Foundation left the active research involvement in telework. The FAST Programme of the European Commission took up the European Community perspective, the results of which are also reported in this conference. In fact, through close cooperation between FAST and the European Foundation, a common momentum of interest was maintained, and the Foundation could return to field research in telework under its 1985/86 programme of work, building on FAST results. In the following chapters, the main findings of two of the Foundation's studies of 1985/86 are briefly summarized.

3. TELEWORK - ATTITUDES OF THE SOCIAL PARTNERS, THE LABOUR FORCE AND POTENTIAL FOR DECENTRALIZED ELECTRONIC WORK IN THE EUROPEAN COMMUNITY.

This project was carried out by empirica for the European Foundation [2]. As empirica are also the organizers of this conference, it is particularly appropriate to relate these findings here.

3.1. Parameters for a definition of telework

As far as numbers of teleworkers are concerned, the situation does not seem to have changed since 1982/83, as there are still only a few thousand people for whom tele-working from home or satellite/neighbourhood offices "is a full-time substitute for working in traditional office environments" [3], as stated in the interim report.

In order to help standardize the discussion about telework, empirica developed the following definitions of telework [4]

- **Location**
  The location of the work site is determined by the needs of the teleworkers and is relocateable as desired or needed. This implies that the geographical site at which work is completed is independent of the employer and/or contractor.

- **Use of Electronic Equipment**
  Telework relies primarily or to a large extent on the use of electronic equipment (PC, storage typewriter etc.).

- **Communication Link to Employer/Contractor,** whereby two different conceptions are used:

  - **Distance Working - Narrow Conception**
    A communication link exists between/among the teleworker and the employer/contractor which is used for electronic communication and transmission of work results.

  - **Distance Working - Broad Conception**
    The teleworker works at a distance (spatially separate) from the employer and/or contractor whereby work results are stored on a disc, cassette etc. There is no electronic communication link used for data transmission. The work results are delivered by traditional media, such as mail, courier, etc.

In their analysis, empirica employed the organisational concepts that were developed by some of the pioneers of telework thinking, such as Diebold Inc. or Margarete H. Olson, who is among the speakers of this conference. Based on this seminal work, the following organisational forms can be discerned:

- Satellite Work Centres
- Neighbourhood Work Centres
- Flexible Work Arrangements
- Work at Home.

Diebold's and Olsons's work can be seen as the first extensive and successful attempt to cover all the various organizational forms that could occur in the framework of telework. Their system proved to be useful and was later adopted by experts all over the world.

Remote information technology based work options are as follows:

- **Satellite Work Centres** are relatively self-contained organizational divisions in one firm physically relocated and separated from the parent firms. The emphasis is on locating these centres within a convenient commuting distance of the greatest number of employees utilizing the site. The number of employees working in a satellite work centre is determined by:

  - "economies of scale", of equipment and services,
  - the maintenance of a sufficient hierachical structure for adequate management on site and
  - sufficient social interaction among employees.

In order to benefit from economies of scale it may be optimal to relocate an entire function such as accounting or data processing. The supervision of work is generally by management staff on site.

- **Neighbourhood Work Centres:** Neighbourhood work centres are offices equipped and financially supported by different companies or organizations. In these offices, employees of the founding organizations share space and equipment in a location close to their homes. While the number of employees is sufficient to provide necessary social interaction, hierarchical structures are generally lacking and supervision of work is carried out remotely.

  One of the principal motivations behind this concept is to reduce employees' time and expense of commuting. In addition, they enable firms to make use of lower office rent outside central cities. These centres, however, are often difficult to implement - particularly on a large scale - as they require extensive cooperation among different organizations.

- **Flexible Work Arrangements** provide employees with flexibility in the scheduling and location of work. This option recognizes the need for occasional alternative work arrangements, especially for professional and managerial employees and provides mechanisms to accommodate staff/family as well as work responsibilities. Furthermore it enables, for instance, DP professionals to accomplish critical work at "non-peak" computer hours or just for convenience. Within this context work forms like "job sharing" and "flexitime" are relevant, too.

- **Work at Home** also labelled "electronic homework" is the most decentralized form of remote work where employees work at home on a regular basis. Under this option, an employee's work week may range anywhere from a few hours to full-time. While homework depends virtually completely on remote supervision and does not provide a field for work related social interaction, it does offer employees maximum flexibility in scheduling working time. In this way, employees may work when and where is most convenient to them.

  Generally work at home can be utilized as an option on an individual basis to accommodate a particular situation or need, either temporarily or permanently. It is also the form of telework which is most frequently utilized by self employed electronic workers.

- **Electronic Service Offices:** These are independent firms which either carry out a wide range of data-processing and computer-related services (word-processing, data-processing, book-keeping, administration of stock or highly specialized work such as secretarial offices which only offer text processing). These firms offer their services primarily to small and medium-sized firms for whom the purchase of a computer appears too expensive. Larger companies may also make use of such services during times when they are experiencing internal bottlenecks or peaks.

3.2. Opinions of the labour force towards telework possibilities

The empirical data which form the basis of the analysis of future potential for telework was obtained in two surveys, an employed persons survey and a decision-maker survey, which were part-financed by the European Foundation together with others, such as the Commission's FAST Programme. Although there seems to be a general potential for decentralisation of certain activities, it is difficult if not impossible to extrapolate from subjective acceptance to general practicability, but it seems at present that roughly 25 % of those whose work could be decentralized would be happy with such a development. Major variations occur with regard to the level of education and with regard to the economic branches of the companies and their respective employees. Above average interest in telework is being shown by EDP specialists, engineers, secretaries/typists, clerical workers, accountants, lawyers and journalists, as well as commercial sales persons; this latter category has, however, a long tradition of decentralized work as "travelling salesmen".

As far as specific branches with the greatest potential need and utilization rate are concerned, banking and insurance, wholesale and retail trade, the private service sector and to some extent the public service sector as well as freelance professionals stand out from the rest. For small and medium-sized enterprises (SMEs) there is a need for external specialist services such as financial planning, payrolling, taxation services etc.

Computer programming and other special EDP activities are also of interest to SMEs in particular, whereas larger companies tend to keep these functions in house. However, decision-makers seem to perceive a host of inhibiting factors against the rapid and/or widespread diffusion of telework. Among the major ones documented are:

- a general scepticism against the technology and against the underlying organisational principle of decentralisation

- organisational complexity and effort

- cost implications of hardware, software and networks

- management and control of personnel and performance level.

It is impossible on the basis of the existing data to determine to what extent these inhibiting opinions are due to the unfamiliarity and lack of experience among the present generation of decision-makers or to what extent they represent a groundswell of opposition and desinterest.

Whatever may be the case, it is possible to deduct - on the basis of supply and demand aspects - that there are potentials for electronic decentralizations of work. There are transnational mass markets for text processing and data input activities. There are also special activities markets such as programming, accounting, etc. which are suitable for decentralization, especially for small and medium sized companies or free lance professionals (architects, doctors, etc.).

3.3. Opinions of the Social Partners Regarding Telework

Trade Unions and Employers' organisation both exhibit a reserved attitude towards the future development of telework and its desirability or otherwise. The greatest reservation is expressed by the unions against one specific form of telework, i.e. electronic home work.

Their worry focuses primarily on the situation of weak persons in the labour market, such as homeworking women with low job content, little or no chance for improvement of skills. They may subsequently become even more marginalized which is obviously an undesirable situation for the persons concerned and for the trade unions who want to protect them from such perils. The general trade union opinion is towards trying to prevent the widespread development of such "unprotected" telehomework places.

In contrast, the employers' federations exhibit a positive disposition towards the idea of work decentralization, although they are sceptical as to the potential to actively promote this under organisational, communicational and control aspects.

These reservations concur with the Decision-Maker-Survey, so that one can generally conclude that the present reality is not conducive to a rapid and widespread diffusion of telework within the general context of the working world. Considering that telehomework as an extreme form of telework seems to arouse most fears and reservations both among potential and actual teleworkers as well as on the trade union side, it seems worthwhile to take a closer look at that particular form of telework and discuss the major arguments for and against based on in-depth case analyses in a number of European Community countries, which were sponsored by the European Foundation. The two substantive aspects under which this work is presented are: to what extent is the psychological space we call "home" suited as a tele-workplace, and secondly, what are the special interests of women in relation to telehomework.

4. ELECTRONIC HOMEWORK AND LIVING ENVIRONMENTS

This chapter relates to work conducted for the European Foundation by the Irish Foundation for Human Development under the guidance of Jean Tansey and Rosalyn Moran [5]. It looks at the likely consequences of electronic homework on the environmental functioning of the workers at the level of the home, neighbourhood and macro-environment. This is an area of concern which is not widely considered in the literature up to now. For this reason, we feel it justified to report in some detail about this perspective and the electronic homeworkers' perceptions of it.

4.1 The Work Room - Some Considerations from Architectural Psychology

Much research has shown the importance of the work environment for workers' performance, health and well-being. Almost all of this research has been carried out in office work environments and environmental aspects of space use have also been well documented in particular in relation to office buildings. Little is known, however, about the impacts of using the home as a workplace.

The physical attributes of the home and the workroom in particular need to accommodate telework. Design, layout, lighting, furnishings, etc., need to suit the needs of the worker. Established activity pattern in the use of space in the home may need to be changed as well as the expectations of people sharing such spaces. The status afforded the home-based work by household members is critically important in this regard.

Overlap of space usage can create conflict and be a major source of stress. When overlap of usage occurs work activities are likely to be disturbed. The extent of experienced disturbance depends on a range of factors, e.g., personal characteristics, nature of the work, required accuracy levels, etc.

If the workspace is not defined architecturally with clear boundaries which separate it from living space, then home workers frequently report having difficulty separating themselves psychologically from work and feel constantly that they should or could be working.

The concept of privacy is relevant to these issues, and is of prime importance in planning workspaces in the home. The concept of privacy relates equally to the three theoretical elements of environmental experience described above. The following aspects of privacy are relevant in relation to telework:

- Psychological privacy, i.e., the ability to regulate social contact.

- Conversational privacy, i.e., the ability to talk without being overheard and the ability to block out surrounding conversations.

- Visual privacy, i.e., the ability to control what one sees and the extent to which one is seen.

- Architectural privacy, i.e., the availability of visual and acoustic barriers.

- Distractions, i.e., background noises and movements.

- Disturbances, i.e., unwanted interruptions.

Privacy can be achieved in a number of ways - through rules, manners, time scheduling, physical barriers, spatial organisation, etc.

4.2 The Home: Some Considerations from Architectural Psychology

One of the difficulties in using the home as a work environment relates to our contemporary Western concept of the home. The home is seen as primary territory and as an extension of the self. The meaning of home emerges from the interaction of opposites, e.g., public vs. private, others vs. self, accessibility vs. inaccessibility. The home thus has deep psychological and socio-cultural significance.

It is likely that doing paid work at home may impact on the atmosphere of the home and hence people's cognitions of the home. Such activities as communicating with one another; being accessible to one another; relaxing after having finished work; being free to do what one wants; being occupied rather than bored, are likely to be influenced by the presence of someone carrying out paid work in the home. This impact is likely to be negative if the physical attributes of the home do not support work activities, or if the expectations or cognitions of household members regarding space use and accessibility are not in accord with those of the homeworker.

The home is seen as shared family space and while a wide range of family related activities are expected and accepted, paid work is usually not. Thus, the person adopting this role in the home is likely to find that social behaviours and norms within the family/household are not supportive. When this happens, the teleworker is likely to experience conflict and stress. On the other hand if the household norms are supportive the home is likely to be seen in a more positive light - perhaps as more interesting, family oriented, etc.

4.3 Residential Amenity and Telework

In addition to the issues raised by telework for the individual's relationship with his/her home, planning policy in relation to the use of the home as a workplace warrants attention and is briefly discussed here.

Most European countries impose restrictions on the use of the home for paid work. The nature and extent of those restrictions vary both within and between countries. Planning policy in this area has traditionally been concerned with protecting residential amenity in the face of industrial production. The changing nature of work in the information society make these concerns less relevant. Many small scale service activities are environmentally benign, yet rigid zoning restricts the development of these types of economic activity in residential areas.

Frequently at the start-up phase of a business, the savings in overheads which result from use of the home as a workplace may be considerable. In addition, for those who prefer to work at home, zoning policies can hinder their efforts. Information supplied by the research coordinators in the participating countries indicated that for the most part the teleworkers surveyed were not unduly concerned about any planning restrictions which operated as they regarded their business activities as "low profile" activities.

In the context of home-based telework, many of the conventional threats to amenity do not apply. It is unlikely to generate noise or air pollution. Possibly, the greatest threat to residential amenity in the case of telework would be the generation of business-related traffic.

4.4 Summary of findings

The sixty-two telehomeworkers covered in this survey, come from Denmark (15), Germany (30), Greece (22), Ireland (2), and the Netherlands (3). With regard to the type of work they are performing, the distribution is as follows:

- Word Processing (usually with one or more of a
  variety of activities, e.g., translating documents)   31
- Programming and Data Analysis   14
- Decision Support   4
- Photocompositing   4
- Accounting   3
- Computer Aided Design   3
- Information Broker   2
- Software Advisor   1
- Total   62

With regard to the "home", the 62 telehomeworkers in the present study live in larger than average accomodation than the national average for their countries. The majority live in apartments but a high percentage live in houses. Most (53%) carried out their work in "an office", 31% in living areas. These workrooms were rarely used by others and respondents reported that they were rarely disturbed while working. The major source of disturbance was the telephone or other household members. Teleworkers were generally satisfied with a range of features of their workrooms, e.g., lighting, furnishing, fittings, etc., and some perceived their home as a pleasant place to work.

The majority of respondents felt it was very important to have a room which could be used solely for their work. For the majority of respondents, their cognitions and feelings about their homes had changed little since they started telework. Full-time teleworkers, however, divided into two groups - those who felt their homes had become less stressful since they took up telework and those who felt their home had become decidedly more stressful. This latter group tended to be females who would prefer to work outside the home.

While the majority of teleworkers reported having less free time since they took up telework, for some, there was a consistent pattern for more or their free time to be spent actually in the home. However, a significant number of those whose only work location was in the home and those who were full-time teleworkers, also spent significantly less of their free time in their home since starting telework. It emerged that these were mostly living in apartments as opposed to houses.

This has a number of Implications for Housing Design : it is clear that the vast majority of respondents felt that the optimal workspace should be clearly defined architecturally - preferably it should be a separate room with a door which could be closed or left open at will. The availability of a separate space is likely to be problematic in small or overcrowded houses. Potential teleworkers need to be aware of the spatial, psychosocial and task requirements of home-based electronic work. If a separate workroom is unavailable, teleworkers may need to learn how to establish privacy by other means, e.g., erecting physical barriers, partitions, creating rules regarding the use of space, time scheduling, etc.

Architectural enclosure is important because it :

- Can maximise visual, acoustic and conversational privacy and minimise disturbances and distractions.
- Allows for the physical separation of work from living space – most respondents felt it was important to be able to get away from work and disengage.

Some respondents wanted to have visual and/or auditory access to other household members while working, others preferred to be inaccessible. Some respondents in the present study felt a satisfactory compromise was reached by leaving their workroom door open while working, and keeping it closed when they were not working. Floorplans should reflect the relationship between living and working spaces desired by the teleworkers.

A related concern is the frequently reported isolation felt by the homeworkers surveyed. This suggests that workrooms for those who experience isolation would be best placed to the front of the house where they are more likely to have access to stimulating views, e.g., to a lively street vs. quiet backgarden. The former view would provide what some teleworkers reported missing greatly, i.e., some sense of rhythm of the day and a feeling of being in tune with the activitiy patterns of others. Research has shown the beneficial effects of having windows and a view into the distance from one's workplace.

The use of the home as a workplace changes the public-private dialectic of the home, particularly if the worker has clients/employers or their agents visiting. A semi-public space may need to be provided to receive visitors such that the worker and his/her household members do not feel their space is being intruded upon.

Such spaces may need to have the characteristics of "front" vs. "back" space – i.e., they may need to be tidy, clean and free from highly personalised and private artifacts. Workers may want such semi-public spaces to convey a specific image, e.g., successful business.

For the on-line telehomeworkers surveyed the telephone was often a source of conflict and stress if they did not have more than one telephone line. Separate lines are advisable for a) receipt of personal calls and for use by other household members and b) making or receiving business calls. The phone was a major source of disturbance for telehomeworkers in the present study, thus the auditory accessibility of non-business phones should be minimised.

There is some evidence that activity spaces may shrink when people change to telehomework. For example, the majority of workers reported visiting the city less often since taking up telehomework. Some commentators have argued that the negative effects of shrinking activity spaces would be compensated for by increased neighbourhood involvement on the part of home based workers and could lead to regeneration of community living.

The results of the study confirm the importance of the journey to work as a multipurpose trip. Respondents use or did use the journey to work and the area around the workplace to meet friends and have recreation. With the reduction or elimination of the journey to work associated environmnental behaviours are likely to be displaced. Uptake of dispersed working arrangements to any significant extent would thus have major implications for land use planning and the provision of facilities.

5. ELECTRONIC HOMEWORK AND WOMEN

The information for this chapter is based on research for the same project as the previous chapter (see note 5). From the outset in 1982, the Foundation's research on telework has been developed with the specific aspect of impact on women in mind. In this respect, the long-standing cooperation between the European Foundation and the relevant services of the Commission of the European Communities, Section "Employment and Equal Treatment for Men and Women" in the Directorate General for Employment, Social Affairs and Education (DG V), has been very fruitful.

The general themes, considered in a context which recognises the interdependence between both working and living experience, are:

a) Women's and men's participation in paid employment (specifically, telework)
b) The division of labour in the home
c) The extent of change in "free time".

This chapter seeks to evaluate the specific impact of telework on these areas both in terms of what has taken place and of potential for future development.

Recognising the difficulties in formulating work/family arrangements, especially for couples with children, which allow both partners to participate in paid employment and to share family life and responsibilities, the Foundation's project considers the potential of telework, deriving from its flexibility as to time and place, for the reorganisation of work so as to provide for such arrangements [6].

While the project has primary regard to the difficulties encountered by women, particularly in combining paid employment and family life, its perspective includes both women and men. It seeks to explore their perceptions of the desirability of telework and of its actual or expected effect on equality in the labour market and in the home.

In examining the labour and domestic roles of those who engage in electronic homework, the project attempts to determine the extent of change in such roles and, in particular, to isolate those changes which are directly attributable to telework **per se.** The difficulties inherent in this are recognised given the multiple and overlapping factors which influence lifestyles - social and economic background, physical location, stage in lifecycle, etc.

5.1 Multiple Roles - One Space

For many, the main implications of telehomeworking are that it results in the home and work space being one and the same. Women who work often have to juggle the conflicting demands of home, husband, children and work. If that work is to be based in the home - the locus of the other roles - the consequences are far-reaching.

Role conflict results from the fact that multiple roles (with corresponding demands and expectations) operate at the same time. It is thus possible that working at home, since it overlaps in time and space with the roles of employee, mother, housekeeper and wife could lead to greater conflict. At present, men show few problems with role conflicts because, unlike women, their roles do not operate simultaneously.

Telework combines home and labour market participation in a unique manner. To understand how telework may develop it is necessary to understand the way in which women partipate in employment, the historical considerations which have tended to influence that participation, the impact of family considerations - especially of motherhood, and the extent to which considerations of equality are altering the patterns of home responsibility and labour market participations.

The extent to which telework is likely to be accepted as a standard option in terms of labour market participation can be explored through seeking to understand the dynamics of home/work relationships and the experiences of those actively engaged in telework.

5.2 Tentative conclusion : Implications for Equality and Wellbeing

The women's aspect of the project has been considered against the background of economic and social changes which have direct implications for women's position in society. It takes telework as one example of new forms of work and activity resulting from technological, economic and social change. Although the study is based on a sample of "only" 62 persons, it centres on "real" people and on their real experiences as teleworkers - rather than being based on hypothetical considerations as some previous work has been. Bearing in mind the limitations imposed, a number of tentative findings and implications emerge, which are broadly summarized in the following list:

- The problem of social isolation and the inhibitions which telehomework places on normal patterns of social interaction are critical areas requiring attention. This is particularly true of people who engage in "telehomework only", who at present and in the predictable future are most likely to be women.

- The dramatic reduction in free time experiences, particularly by women, was significant. It raises the question of what "flexibility" is.

- There was flexibility of place, and the fact that the "place" was home provided telehomework with one of its greatest attractions : it could be combined with childcare, etc. It also resulted in its greatest dangers - role conflict, space not properly geared for work, a blurring of the work/free time division, and, for some, a tendency to overwork. It also reinforced the sense of "isolation" which already exists for many housewives.

- The study confirms the limited degree of change which is occurring with regard to division of labour within the home. There is no hint of a fundamental shift in attitudes or practice, although women were doing slightly less and men were undertaking slightly more housework and childcare.

- Women have different values with regard to children. Many women stayed at home because they had a desire to be with their children and for some it seemed the only way they could combine work and family responsibilities.

- One of the findings was that so many women actually enjoyed working with technology - they found it interesting and challenging. There was an interest by some women in establishing themselves as "self employed", and in using home teleworking as a base upon which to do that. It is possible that in certain instances telehomework at this stage is appealing to people with some entrepreneurial skill and it appears, again within the limited sample, to attract those of higher socio-economic background.

The outcome of the survey underlines the point made earlier that greater awareness of time-use is an essential ingredient in coming to a fuller realisation of the practice and potential of telework. A true account of the actual disposition of time by men and women, at home and at work, before and after adopting telehomework is perhaps the most effective way to evaluate telework's real impact.

There are a number of general policy indications:

- A particularly high percentage of women favoured a mixture of "home and other workplace". Such an arrangement obviously is the ideal for many : it is of course the reality for many men at present. It suggests that for purposes of planning or piloting of future work arrangements a model which allows some outside location as well as the home is one to be explored and encouraged. This suggests that concepts such as the "neighbourhood centre", should be fully explored as part of any development of telework, taking account however, of the need to avoid any possibility of the "ghettoisation" of women in such centres. At this stage, while the literature does refer to them, there are very few examples, and as yet little concrete evidence of success.

- There was a clear-cut finding that participants did not want both partners at home. This has considerable implications for future planning and development of telework. If it is the case that couples prefer not to have both at home working, the future possible scenario of women at home teleworking and men working outside the home is strengthened. Given present day gender relations this seems somewhat more likely than men staying at home teleworking while women go out. Such a development would be likely to reinforce gender role division, and exacerbate the social isolation of women.

- Many of the women respondents referred to health problems, or potential health problems. While specific investigation of health implications are beyond the scope of this study, the findings are such as to suggest that telework will unquestionably require monitoring with the utmost care and precautions of the most rigorous nature in this area. It raises in a very specific way the problems of monitoring, organising or advising teleworkers.

- The particular nature of electronic homework throws into sharp relief the need for a comprehensive system of statutory protection. The range of legislative supports available for people at work in conventional workplaces does not exist for home-teleworkers and to the extent that legislative provisions exist in some countries their implementation is difficult to achieve. The essentially disparate nature of electronic homework renders it virtually impossible for the workers to organise themselves effectively and this makes them vulnerable in the matter of conditions of work. (Although it is perhaps a consideration for the future, the ability to use technology to communicate and organise may be an advantage to teleworkers).

. The present sample with which we are dealing is representative of higher socio-economic/education groups. Clearly if telework is to develop and become more widespread it will embrace a broader spectrum of people whose domestic facilities will be less than those of the respondents in the present survey, and which may in some instances be scarcely adequate for the ordinary business of living, quite apart from the additional function of telework within the home. In such circumstances all of the problems encountered by respondents in the present study will be expanded - questions of organisation, space, health and conflicting roles.

Present day working arrangements have been far from ideal as far as women are concerned, and employment demands of office or factory have often involved bad working environments and difficult travel or childcare arrangements.

New technology is with us, and it will continue to increase in impact and influence. It provides flexibility which at least potentially, could be used by women in an effort to put forward some ideal in terms of a society in which work and family and children can co-exist comfortably.

6. CONCLUSIONS: A TENTATIVE FORECAST

The diffusion of the various forms of telework is likely to remain rather limited, both within the Member states of the European Community and even more so in cross-national applications. There are a number of reasons for this of a cultural-linguistic, political, organisational, and also to some extent of a technological nature:

- There is no common "economic" language in the European Community. Although the internationally dominant language of the "computer culture" is English (or American), which is also one of the official languages within the Community institutions, business transactions are done in the various languages of each country, with limited use of foreign languages for trans-national activities. Telework configurations covering several countries with different languages are not likely to spread very much as they would depend on a formal or at least informal agreement on a commonly understood language. Considering the low priority of language teaching in most national educational systems, progress on this level will be slow and limited;

- Despite the aspirations of the European Community to strive towards the creation of an internal economic market including services by 1992, such services requiring a common mode of communication (language) and a cross-national tele-communications network are likely to be severely handicapped, as neither of these exist for the moment;

- What exists at the moment in terms of cross-border telecommunications links is not only limited in its capacity as compared to the internal networks, but is further discriminated by higher charges to the user. If the political decision makers would be able to bring about a pattern of telecommunications charges similar to the unified postal charges for a standard letter, i.e. the same piece of information costing the same amount regardless of distance or country of destination within the European Community, a major step towards an internationalization of telecommunications would have been achieved - thus paving the way for a potential spread of international telework;

- Organisational structures, organisational behaviour in general, management styles and cultures, trade union policies and attitudes and opinions of employees all point in one direction: centralized rigidity as an organisational principle is easier to handle and to live with than flexibility and uncertainty in a more decentralized organizational model. The fact that telework does run counter to many of the established "easy-to-live-with" and seemingly well-proven organisational principles, suggests that it will only develop at the same pace as the organisational culture as a whole moves away from centralized rigidity, predictability and control to more flexible - but more difficult to manage-systems. Although "flexibility" is very much under discussion at the moment - with the employers supposedly for it and the unions and employees supposedly against it - its rapid and widespread diffusion as an organisational principle is unlikely, if for no other reason than the inbuilt inertia which existing systems tend to exhibit, not to mention the humans which inhabit such systems;

- the potential technological inhibitions towards a widespread cross-national European telework scene are closely linked to the political inhibitions mentioned earlier. It would be technologically easy to establish the necessary cross-national networks, but it takes national political decisions to bring about such a development. This is, of course, the dilemma of any "European" policy decision which is made under the limitations of "national" interest considerations. It is to be hoped that the active and constructive implementation of the Single European Act, in conjunction with the aim of creating an internal market by 1992, will accelerate the development of European Community oriented policies at all levels of the decision-making process, be it the Commission, the Council of Ministers, the European Parliament, the national parliaments in the Member States, or the two sides of industry at European and national level.

Taking all these general considerations into account, and adding the worries about specific forms of telework, such as electronic home-work, which are insufficiently covered by existing labour law and health and safety regulations, one can see an orderly development of telework only in the context of the mutual interest of employers and employees and trade unions. Public bodies can of course do their part by assuring - through updating of the relevant bodies of law and regulations - that socially undesirable developments are being contained.

There are objective and subjective advantages attached to telework, such as an individualisation of working time and working place, greater possibility for combining several roles such as income, family and leisure, as far as the employed persons are concerned. For employers, the benefits could be seen in more variable deployment of labour, less need for expensive centrally located office space or access to specialist knowledge on a contract basis. But if all of these positive factors can outweigh the deep-seated fears and suspicions towards the "unknown" remains to be seen.

ACKNOWLEDGMENTS

The above analysis is based on quantitative and qualitative data which were collected between 1982/86 in the following Member States of the European Community: Denmark, France, Germany, Greece, Ireland, Italy, the Netherlands and the U.K. The authors of this paper would like to thank foremost all the respondents of the various surveys and interviews that were carried out. Furthermore, the following researchers who worked on the Foundation's projects, merit special mention for their interest and insight into the phenomenon of telework : Sylvie Craipeau, Marylin Dover, Gisela Erler, Monika Jaeckel, Werner B. Korte, Jean-Claude Marot, Daniella Mazzonis, Clio Presvelou, Roberta Shapiro, Wolfgang J. Steinle, Kelly Syrigou, Gitte Vedel.

As usual, any flaws that may exist in the judgment of facts and trends expressed in this paper are the responsibility of the authors alone.

NOTES

1) Telework: Impact on Living and Working Conditions. Dublin: European Foundation for the Improvement of Living and Working Conditions, 1985, published in French and English.

2) Telearbeit - Meinungen und Standpunkte der Sozialpartner und der Erwerbstätigen sowie das Potential dezentraler informationstechnisch gestützter Büroarbeit in Europa. Final report for European Foundation for the Improvement of Living and Working Conditions, Dublin, October 1986. Unpublished typescript in German.

3) Interim report of 2) above, April 1986.

4) Interim report of 2) above, p.3.

5) Telework - Women and Environments. Final report for the European Foundation for the Improvement of Living and Working Conditions, Dublin, November 1986. Unpublished typescript in English.

6) Tentative proof of a shift in activities over the last 25 years or so, i.e., women doing less housework, men doing more, is contained in another project which the European Foundation has undertaken in the past few years : the establishment of the "European Foundation Archive of International Time Budget Data", in conjunction with Prof. Jonathan Gershuny of the University of Bath. An unpublished interim report details these findings : Time Use in Seven Countries 1961-1984. Report for the European Foundation for the Improvement of Living and Working Conditions, Dublin, November 1986 (in English).

"DISTANCE JOBS" - A NEED FOR EUROPEAN ACTION

WOBBE, Werner

DG XII - FAST
Commission of the European Communities
200 rue de la Loi
B-1049 Brussels
Belgium

"Distance working" is already evident in Europe in a wide variety of forms, although it is a relatively new phenomenon. New Information Technologies (NIT) have reopened the question of where people work. Patterns of commuting to work, the advantages of centralised work places and the location decisions of firms are all now subject to review.

Before a possible broader take-off occurs, the FAST Programme of the Commission of the European Communities* is interested in the potential future implications. The following questions therefore arise:

What new developments, new forms of distance working are expected to emerge in the next 10 - 15 years in connection with the application of new information and communication technologies? Will distance working at home undergo a new growth and expansion? Will it mean a return 'en masse' of women to the home? What kind of occupations could be more directly affected by the different forms of distance working? Should we anticipate important changes in:

- enterprise cooperation
- work organisation
- urban and regional location of activities
- development of peripheral and rural areas
- technological infrastructures?

These questions formed the basis of the FAST study (TWE 6): Distance Working in Urban and Rural Settings. FAST was particularly interested in knowing how present R & D activities will be affected and what could and should be the role of the Commission of the European Communities in promoting the optimum use of such new technological opportunities?

The FAST study was a cross-national research enterprise involving the Tavistock Institute (UK), Association d'Etudes et d'Aide pour le Développement Rural (France) and Empirica GmbH (Germany). To accelerate communication, the research teams and FAST utilised new forms of distance working equipment - electronic mail and conferencing - together with the telefax system, and assessed their value for research practices. In addition to the three participating research teams, further contributions have been added to the FAST network from Belgium, France, Germany, Ireland and Italy.**

The FAST DISTANCE WORKING STUDY was organised around five research activities.

1. As a first step, experiments and experiences in distance working projects in the Federal Republic of Germany, France and the United Kingdom were studied. 43 projects were reviewed in total. This activity was conducted by cooperating research teams from three countries - Empirica (D), ADR (F) and Tavistock (UK).

2. The second step was to define distance working today. Investigatory work based on the experiences arising from projects studied focused on telematic based developments. "Distance Working" was chosen as a label to cover many different forms. This activity was undertaken by Tavistock (UK) and resulted in the following classification:

   - Homework is recognised as paid employment undertaken by a person working entirely or to the most part at home, with visits to the site of the employer or client.

   - Shared facilities are offices or work centres which are equipped with various electronic facilities. The offices are shared by a number of users, who are unable to afford such facilities of their own. They also offer additional opportunities to those employed, such as training schemes, childcare facilities, conference suites, etc. In this form the neighbourhood centres are know mostly as suburban facilities. However, they are a sub category like metropolisation and regionally based shared facilities.

   - Distance working enterprises are the third category. This enterprise offers new information technology based services to customers located at a distance, like one-line data base access, accountancy or document preparation. In this form, for example, distance working enterprises could be established for disabled people.

   - In satellite/branch offices, enterprises relocate part of their operations at a distance from the main body, maintaining communication with head office by telematics.

   - Distributed Business Systems is defined as a number of separately located units involved in different stages of the production of an end product or service, linked together by NIT within a total, production, service or distribution system. In special cases, small separate enterprises cooperate to achieve the economics of scale in some special functions of their business.

   - The last form is mobile work, which is very well known to sale representatives or service engineers. They work in more than one place and communicate with headquarters by the use of portable communication facilities.

   These different forms of distance working widen the scope of the discussion on teleworking, which until now has centred largely round home working. it also indicated the broad potential and implications new telematic facilities hold for:

   . organisation and management of firms
   . wages and working conditions of employees
   . career patterns
   . vocational training
   . mobility schemes and implications for families

3. A third step was to outline attitudes towards distance working and their quantitative relevance. The problem in question is:

   Does telework present any potential for new social innovation in view of the users; will it be relevant in the future? A representative survey involving more than 10,000 people in four European countries - the Federal Republic of Germany, France, Italy and the United Kingdom, carried out by Empirica (D), estimated that more than 12 million employees in these countries are interested not only in telework in a broad sense, but also in the more narrow sense of electronic homework. There are important variations between the countries, but the survey does indicate a huge potential for future changes in work and employment conditions.

4. A large part of the study was devoted to the examination of the implications and possible challenges to <u>social and economic development</u>. As part of this activity, national seminars were organised by each of the research teams in France, Germany and the United Kingdom. These considered urban, rural and regional development; implications for particular social groups such as women, old and young people; employment questions and qualifications; the relationship between home and work. The seminars were organised by ADR, Empirica and Tavistock with assistance also from the Carrefour International de Communication (F).

5. Finally, an evaluation was made of the possible <u>conclusions</u> of the work undertaken and <u>recommendations</u> for action to the <u>Commission</u> of the European Communities and its Member States. The conclusions to the research point out that:

- Not all forms will develop automatically, especially those with additional social aims and supporting regional development. Therefore local and regional funding will be necessary, above all for shared facilities with additional services for employees - child care and training facilities, etc. Also distance work enterprises for disabled people have to be supported.

- Regulatory measures have to be developed to guard against possible disadvantages foreseen in working conditions, particularly with regard to homeworkers.

- As regards regional development, special communication lines have to be established and tarifs set, based not on the distance of the transmission, but rather on the quantity to ensure comparable conditions for the user with distant regional sites.

- A huge demand for distance working in the peripheral areas was expressed in the seminars, which does not correspond to the level of qualifications available with regard to NIT facilities, as the empirical survey points out. Therefore, training has to be offered in these regions.

The recommendations which follow from this research refer to pilot projects, diffusion policies, regulatory measures and research and development. Pilot projects, together with R & D and regulatory measures, are considered of particular importance, and could play a significant role in future Community policies. Different organisational forms of distance working within these <u>pilot projects</u> could be applied according to regional requirements.

■ Industrial Centres in Peripheral Regions
   . Shared facilities in regional centres
   . Modernisation for larger regional enterprises

■ Peripheral Rural Areas
   . Rural distance working enterprises
   . Rural shared facilities

■ Deprived Inner City Areas
   . Decentralisation of local administration
   . Inner city shared facilities

These pilot projects should aim at:

- giving access to disadvantaged regions in the labour market
- opening up employment possibilities to handicapped social groups
- fostering new competitive structures of enterprise organisation

A European action in the field of pilot projects could provide:

- assistance in the establishment of pilot projects in Member countries (see the proposed distribution above)
- consultancy in technical, vocational and organisation questions
- comparative evaluation research of the pilot projects
- public display of successful attempts

FAST will examine the possibility of setting up a European plan of action with regard to ERUO-DISTANCE JOBS.

FOOTNOTES

* FAST - Forecasting and Assessment in Science and Technology - is a research programme within Directorate General XII of the Commission of the European Communities.

** For a full list of references to the research outputs of the FAST Distance Working Study see below.

PUBLICATIONS IN THE FRAMEWORK OF THE FAST DISTANCE WORKING PROJECT

(1) Trends and Prospects of Electronic Distance Working - Results of a Survey in the Major European Countries: Empirica, Bonn
Luxembourg 1986, FAST Series No. 20, 60 p.

(2) Distance Working Projects in the Federal Republic of Germany, France and the United Kingdom: ADR, Empirica, Tavistock
Brussels, March 1986, FOP No. 79, 55 p.

(3) R.W. Holti and E.D. Stern: The Origins and Diffusion of Distance Working
Edition Futuribles, Paris 1986: forthcoming, approx. 200 p.

(4) Technical Aspects of Distance Working: T. Fogelman (1985), 25 p.

(5) Reports on Workshops in France, The FRG and the U.K.:
ADR, Empirica, Tavistock
Brussels, September 1986, FOP No. 117, 164 p.

(6) Distance Working Study: Conclusions and Recommendations:
E.D. Stern and R.W. Holti, Brussels 1986, FOP No. 92, 52 p.

(7) R. Moran, J. Tansey: Distance Working: Women and Environments.
Brussels, March 1986, FOP No. 78, 43 p.

(8) M. Goldmann, G. Richter: Telehomework by Women. Results of an Empirical Study in Germany. Brussels, November 1986, FOP No. 120, 29 p.

6

MAJOR THEMES IN THE DISCUSSION OF TELEWORK

MAJOR THEMES IN THE DISCUSSION ON TELEWORK

Simon Robinson

empirica GmbH
Kaiserstr. 29-31
D-5300 Bonn 1

1.   INTRODUCTION

In reporting here on the main points arising out of the contributions to the plenary discussion at the Empirica telework conference, these have been organized into 9 topics:

o the shape of telework now and to come
o telecommunications and the PTTs
o telework and societal trends
o management, organizational and job design issues
o the importance of social issues versus technology issues
o women and telework
o education and training
o telework and regional development
o governmental and European policy options

2.   THE SHAPE OF TELEWORK NOW AND TO COME

There was some controversy as to whether the term 'telework' suffices to describe the phenomena discussed at the conference, or whether a new term is needed. Whereas V.S "Steve" Shirley felt 'telework', meaning simply work at a distance, quite adequately describes the form of work practised in her organization, F International, Roger Walker, ex Rank Xerox, was more inclined to look for other descriptions. In the Rank Xerox scheme teleworkers are referred to as 'networkers'.

Gil Gordon, of Gil Gordon Associates, New Jersey, suggested there is a need not just for a new term for telework but for a new terminology. A whole new set of words is required to describe the many forms of work lying between the status of employees and freelancers.

The discussion of terminological questions tends to cover up a real diversity in the forms of telework practised today. Wolfgang Steinle, Empirica, made the point that many contributors to the conference had concentrated explicitly or implicitly on one category of telework, on home-based telework or 'electronic homework' - to the near exclusion of other forms. Steinle proposed that, in future, more attention should be paid to other forms of telework.

Helmut Drüke of IZT, Berlin, suggested the promotion of particular other forms of telework, specifically of neighbourhood or satellite offices [1]. In contrast to electronic homework, he believed these forms of telework offer enriched job content, face-to-face communication and, important particularly in a German legal context, a form of organization which fits into existing legislative provisions for worker protection.

Apart from neighbourhood offices, recommended by several speakers as an improvement on home-based telework, Professor L.S. Harms, of the University of Hawaii, described progress toward the establishment of international teleworking research groups. These groups are supported by small group conferencing systems and file-transfer software. This kind of development is leading to a form of telework he termed 'networked institutions'.

In the final discussion there was little mention of the limits which bound the possible substitution of telecommunications for personal contact in working relationships. Gerard Blanc, of the Association Internationale Futuribles, did point out that for some forms of communication in work, face-to-face contact was preferable to any currently available electronic tele-contact. Negotiations were one form of communication which found specific mention. However, it can reasonably be argued that the outcome of negotiations cannot be objectively judged, being simply either more or less favourable to each of the parties involved. It became evident that the task-related potential, not only for media substitution, but also for the reorganization of communication patterns in support of telework requires further investigation [2,3].

3. TELECOMMUNICATIONS AND THE PTTs

Several speakers and participants (Wolfgang Heilmann, of INTEGRATA, Ursula Huws, of Empirica UK, Andrew Gillespie of the University of Newcastle, among others) referred to telecommunications costs as constraining telework uptake. Generally, the high costs of telecommunications were seen as hindering the diffusion of telework through society.

There were many references, too, to there being the problem not just of high costs as such, but also of the way the tariffs teleworking organizations face are structured. The telecommunications tariff structure forms a barrier to the spread of telework in that the pricing structure of telecommunications in Europe continues to be strongly distance-related.

This is despite the fact that the cost of electronic services is today far less dependent on the distance of communication than used to be the case. The current practice of strongly linking tariffs for telecommunication connections to the distance covered by the link is today only a viable pricing policy where all electronic networks are in the hands of a single monopolist. The other services competing in this communications market - postal services or road or rail transport - continue to have costs, and therefore prices, strongly related to distance.

The fact that tariff structures now unnecessarily penalize long-distance telecommunications must be seen as constraining telework uptake [4,5]. It was suggested that ISDN will not only not solve everything but that this service will take some time to take effect. Tariff structures of current services could be revised, and until they are, one of the major positive benefits of telework, impetus to regional development, continues to be seriously hindered.

Though deregulation and the creation of effective competition in the telecommunications services provision can be expected to force providers to charge prices closer to costs, deregulation and market forces will not lead to purely desirable results. Connection costs are related to the geographical distribution of the network connections, so that uncontrolled deregulation of service charges would tend to lead to discrimination against thinly populated areas. This was pointed to by one contributor to the discussion as yet another factor discriminating in favour of densely populated areas, potentially resulting in a new division in society between the communications rich and the communications poor.

Huws proposed that a tariff structure entirely unrelated to distance ought to be seriously considered, citing the example of the historic introduction of the the Penny Post in Britain on the initiative of Roland Hill. She suggested that there is a need for a Roland Hill of telecommunications. Tariffs unrelated to distance would considerably improve the viability of remote work, thus allowing the benefits of telework to be exploited. Blanc for one considered this an area where the European Community ought to take action.

The structure of services provided by European PTTs is relevant to telework viability and hence diffusion in another respect: there is the question of standardization. Shirley called for European action to achieve more compatibility between national systems. Heilmann also stressed the technical deficiencies of current service provision for effective communication links in working relations.

4. TELEWORK AND SOCIETAL TRENDS

It was made clear by several speakers during the conference, especially by Huws, that the development of telework cannot be seen in isolation from several major societal trends. Werner Wobbe of the FAST Programme of the Commission of the European Communities described some facets of these trends, here supplemented by points made by Marilyn Mehlman of Delfi Consult AB, Sweden. A decline of mass production systems can generally be observed, together with a decline in the prevalence of Taylorized work sequences for which skills are easily acquired. The importance of the trend toward knowledge-based work, production systems and economies and away from land- and capital-based systems also received comment.

Given the move to information and service based economies, new concepts of work organization are in demand.

A major trend in organizational policy towards an expanding use of subcontracting to perform service functions required by organizations in the economy was described by Huws and by Professor Herbert Kubicek of the University of Trier. Typical structural change in organizations involves merging and growth, but there is at the same time a tendency for organizations to get 'ragged at the edges', shifting work onto those previously clearly outside the organization. A further example is the way customers are increasingly required to 'work' to serve themselves. There was general agreement that there will continue to be a trend toward more self-servicing by consumers in the economy.

Huws, while recognizing that changes and the concomitant casualization of work give rise to new challenges and opportunities, also pointed out that while opportunities accrue to some, there is the danger that other groups will tend to face more problems. Some telework schemes have to be seen as casualization in this sense.

5. MANAGEMENT, ORGANIZATIONAL AND JOB DESIGN ISSUES

In some ways the large organization must be seen as the antithesis of teleworking. Telework units tend to be small scale and flexible, especially in the autonomous, entrepreneurial form - as for instance practised by Rank Xerox.

Teleworking pitches small size and flexibility against the inflexible bureaucratic hugeness of today's multi-national corporations.

Margrethe Olson, New York University Graduate School of Business Administration, reported that, though mega-corporations in the USA might still be alive, they are far from being well. For structural reasons, large corporations move very slowly and tend to be poor at adapting. Several speakers, including Shirley, pointed to the high inertia of corporate management as hindering the growth of new and flexible forms of working organization such as telework. Olson, referring to a crisis in corporate management, suggested the tools available to cope with changing environments are technology and human resources: management is too often failing to achieve fit between these two areas. She declared that, in part because of management failings, we have as yet barely scratched the surface of technology and telework potential.

In response to changing market environments, management strategies are beginning to change, and so are organizations. There were several references in the discussion to the ways organizations are attempting to adapt to increasingly unstable environments. Though many continue to grow in size, corporations are changing in structure. There is emphasis today less on earnings to capital ratios and more on "strength to weight" ratios such as given by the ratio of core staff to overall production volume [6]. This is a symptom of increased recognition of the need for flexibility in market response, and of the need for a measure of the achievement of flexibility. Just-in-time production is currently a buzzword in management circles, and it was suggested in discussion that telework has a decisive role to play in enabling this cost-cutting response to increased market volatility. This and other ways in which flexibilization of organization is achieved, such as intrapreneurship, the establishment of profit centres or the externalization of service provision, all tend to point in the way of an increasing role for telework.

The discussion was not limited to the structural inertia of organizations, but also addressed the issue of management competence in more general terms. One area of management of particular relevance is tele-management. Exercising management functions at the end of a telecommunications link requires new skills. Harms for one saw the need for special post-graduate training of management to provide these skills, and reported being engaged in setting up a masters degree course in tele-management. Barbara Klein, of the Fraunhofer Institut IAO, went beyond management training in calling for the development of organizational forms specifically designed to cope with telework.

Mehlman underlined the importance of managers' role in future development, insisting that the key to management in future is uncertainty: management needs to be taught both to live with and to manage uncertainty.

Related to this is an area of organizational management concerned with the design of organizational structures and of jobs in organizations. The tendency is to limit the search for solutions to those modes of working already known, and there were calls that this tendency should be resisted. Constance Perrin of MIT agreed that in future, there will (have to) be more concern with job design in general. The issue will be desirable and undesirable work, whether telework or not - it being undesirable in any context for work to be designed, for instance, with no form of personal development opportunities or career progression.

Speaking about the way she saw future development, Mehlman suggested that the key element is a concept of 'quality'. She warned against limiting the idea of quality to purely technical quality, arguing that the issue is the quality of the whole of working life. Having specified what makes a good job and what makes a good organization, the task in telework is, just as elsewhere, to design quality jobs in quality organizations.

There was some mention of the changing role of work in modern society, questions being asked by Steven Johnson, University of Newcastle, as to whether it is still appropriate to assume that work is as central to life as has been shown to be the case in the past for many of the employed. If the role of work has changed, being diminished in importance, the question is raised as to why work should be the place where isolation is combated (given that a major disadvantage of home-based telework is the social isolation suffered by some teleworkers). Where Johnson was suggesting the diminishing importance of the work place for social needs, Mehlmann argued that, given the rise in one-person households, it might well be that this group at least would tend to move its "territorial base" from the home to the work place.

6. THE IMPORTANCE OF SOCIAL ISSUES VERSUS TECHNOLOGY ISSUES

Roger Walker for one pointed out that technology is secondary in the development of telework. Telework implementations need to focus on the needs of the individual and the family unit.

Bernd Kramer of the German Federal Ministry for Research and Technology (BMFT) picked this up, summarizing speakers' opinions up to then as generally expressing the attitude that technology must be seen as a constraint to development. This drew a rejoinder from Olsen, who pointed to the expanding capability of available technology as being a real motor of development and that there were many current technological options as yet unexploited.

Gerard Blanc found that the European research programmes tended to concentrate on the promotion of technology, of software and hardware. He missed research on the social aspects of the use of technology. Blanc proposed more expenditure of effort in developing software better related to needs and more enquiry into the why of a technology, what needs it meets and for whom.

This was the main thrust of Herbert Kubicek's contribution to the conference and, in the discussion, Kubicek argued again against the tendency to neglect social planning. Political consideration of social issues and the implementation of the ensuing planning requirements tended to lag technology development and implementation. The danger is, he pointed out, that changes in production technology and organization, such as the trend to just-in-time production, will take place without an equivalent effort being invested in adapting the social relations affected by the change.

Huws argued that changes in social structure is needed to make telework a real choice for everyone. Because of the inappropriateness of current social institutions, few have the opportunity to choose whether to take up telework.

7. WOMEN AND TELEWORK

Telework, which is to be seen as part of a process of societal change, was seen by some speakers to affect women in the first instance. Barbara Klein called for more protection for women in the labour market. Some controversy developed on the role of women in telework today and in the future.

Johnson suggested that surveys on telework potential have tended to ask the wrong question to the wrong people, in that most often it is an organization which is the object of investigation. He thought it advisable instead to ask women with young children their attitudes.

It should be added here that only recently has any empirical evidence on the potential for telework in respect of interest in telework in the population been gathered. Most surveys have covered practising teleworkers and not potential teleworkers [7].

Huws reported a high level of interest among the general public in telework, recounting her experience that any public mention made of home-based telework results in a flood of callers interested in the opportunities.

The view that home-based telework is a good opportunity to combine work with child-care is quite common, and this was more or less explicitly expressed in some contributions to the discussion. However, the opinion that the greatest potential for telework lay in that segment of the population formed by women with small children was generally rejected, among others by Blanc, Olsen and Huws. Views expressed here were that home-based telework is no alternative to the provision of child-care. Child-care and work are together two full-time jobs. Women with small children are not, therefore, a natural population for telework.

8. EDUCATION AND TRAINING

The expansion of teleworking has implications for education and training, as several speakers pointed out in different contexts. The need for management training has already been mentioned.

Eberhard Köhler of the European Foundation for the Improvement of Living and Working Conditions, Dublin, pointed out that telework tended to bring more choice and more flexibility to working arrangements. However, flexibility also tends to result in uncertainty. New skills, in particular decision making and responsibility skills, are needed to deal with this. He suggested that societies are apparently not ready to teach these skills.

There is the general point that the trend to knowledge-based economies demand an well educated and trained work force for their functioning. One specific area, computer literacy, was also mentioned repeatedly as an important deficit today hindering the implementation of telework.

9. TELEWORK AND REGIONAL DEVELOPMENT

Several contributors pointed to the potential of telework, i.e. the use of IT and telecommunications to support working relationships, to assist in problems of regional development and in combating inner city decay. Telework was widely seen as offering considerable opportunities to aid underdeveloped or decayed regions.

Shirley and Köhler responded that telework was more than just an opportunity to aid poorer regions, the fact being that appropriate schemes were already having an effect. F International provides many jobs in depressed areas, and Shirley stressed that this choice was for commercial rather than philanthropic reasons. Köhler described the case of a US company which has for some years provided employment in a satellite office in Ireland.

Blanc underlined the local policy issues which arise as decentralization through telework spreads, one being that local administration becomes more important. As Johnson pointed out, decentralizing knowledge or work is not the same as the decentralization of power, and he pleaded for more decentralization of this latter kind. The UK and France are particularly weak in the devolution of power to regions, the UK in particular showing the reverse trend of the centralization of power in the last few years.

10. GOVERNMENTAL AND EUROPEAN POLICY OPTIONS

Governmental policy and European policy issues arise in respect of points already mentioned. The following summarizes other points made in the discussion which directly bear on future policy issues.

o National differences in legislation on data security were seen to be a problem. Riccardo Petrella, FAST Programme, Commission of the European Communities, pointed to action already being taken at European level.

- Legislative effort was seen to be required to counter the danger of groups of workers dropping out of institutional forms of collective security such as insurance schemes. A potential threat to trade union effectiveness had also to be considered and acted on.

- Deregulation policy:
  Some differing opinions came to light on the subject of regulation and deregulation applying to telework. There was a generally observable difference between those speakers and delegates responsible for the running of profitable telework schemes in practice on the one hand, compared on the other hand with those with less direct interest in the viability of individual types of telework schemes. The former, the practitioners, were generally in favour of a deregulation of services and of work organization.

  Heilmann, for instance, who manages an organization having several teleprogrammers, pleaded for free market forces to be allowed to shape the organization of work, that is for the legal regulation of work practices to be weakened. He argued that this freedom would help new structures emerge, in social areas too. Shirley, of F International, also desired more flexible regulations, having, for example, found the provisions of planning laws contrary to the implementation of a teleworking organization. Another point in this direction was made by Walker, in this case that small businessmen are stifled by legislation and by the attention of tax authorities driven by a suspicion that wealth-creation might be involved.

  Those with responsibility for guiding European policy on the issues germane to telework took a somewhat different view. Petrella and Wobbe rejected a laissez-faire, free market forces approach, arguing that analysis of what might happen is not enough: we know what would happen if nothing were done, and since that would not be acceptable, effort is required to actively shape the future.

- There was a call for more application of research findings (e.g. by Mehlman) and for more research generally (e.g. by Blanc), drawing the response from Petrella that there was now already a great deal of research in progress.

- Wobbe described initiatives already started by the FAST programme for implementing special projects related to telework. These include Euro-distance-jobs, providing access to the labour market for depressed regions and weak groups such as the disabled.

- It was stressed that technology must have a useful end for it to be used, and further effort is necessary in the search for useful applications.

- Steinle defended the EEC ESPRIT programme as not being exclusively concerned with technological developments but examining both the use of technology and its social implications. There is, of course, general acceptance that any implementation of technological innovations without regard to their use or their implications for the groups affected will risk failure and costly abandonment. Steinle pleaded for more cooperation of technologists and social scientists in this field.

A major theme throughout the conference was that the potential for telework which undoubtably exists presents both chances and risks. Speakers differed somewhat in the emphasis they placed on each. It is hoped that the points made by contributors to the discussion at the Empirica conference will enable us to achieve a better understanding of both chances and risks to the different groups in societies, an understanding which can be used actively in guiding policy-making at all levels.

REFERENCES

[1]   On the various forms of telework, see the paper by Werner B. Korte in this publication.

[2]   Goddard, J.B./Morris, D., 1979: The Communications Factor in Office Decentralization. In: Diamond, D.R./Mcloughlin, J.B. (eds.): Progress in Planning, Vol. 6, pp. 1-80. Oxford

[3]   empirica 1987: Work Content and Tasks of Telework. ESPRIT project. empirica working paper no. 10. Bonn

[4]   empirica 1986: The Telecommunications Infrastructure Lending Itself to Telework in the Federal Republic of Germany, France, Italy and the United Kingdom - Availability, Applications and Costs (TECOM). ESPRIT project 1030, empirica working paper no. 5, Bonn

[5]   empirica 1987: Overall Report: Potential and Uptake Dynamics of Telework. ESPRIT project. Bonn

[6]   Michael Dixon: "Learning from the Dinosaurs". In: Financial Times, 16.7.87

[7]   Though relevant surveys with all women or the general public as its sampling universe is not known to the author, Empirica has itself carried out the largest survey of potential teleworkers in four European countries, based on the universe of all employed people. Cf. empirica 1987: Profiles of the Population Potentially Concerned with Telework - The Supply of Teleworkers. Results of the Employed People Survey (EPS). ESPRIT project 1030, empirica working paper no. 6, Bonn